St Antony's Series

General Editor: **Jan Zielonka** (2004–), Fellow of St Antony's College, Oxford

Recent titles include:

Lotte Hughes
MOVING THE MAASAI
A Colonial Misadventure

Fiona Macaulay
GENDER POLITICS IN BRAZIL AND CHILE
The Role of Parties in National and Local Policymaking

Stephen Whitefield *(editor)*
POLITICAL CULTURE AND POST-COMMUNISM

José Esteban Castro
WATER, POWER AND CITIZENSHIP
Social Struggle in the Basin of Mexico

Valpy FitzGerald and Rosemary Thorp *(editors)*
ECONOMIC DOCTRINES IN LATIN AMERICA
Origins, Embedding and Evolution

Victoria D. Alexander and Marilyn Rueschemeyer
ART AND THE STATE
The Visual Arts in Comparative Perspective

Ailish Johnson
EUROPEAN WELFARE STATES AND SUPRANATIONAL GOVERNANCE OF
SOCIAL POLICY

Archie Brown *(editor)*
THE DEMISE OF MARXISM-LENINISM IN RUSSIA

Thomas Boghardt
SPIES OF THE KAISER
German Covert Operations in Great Britain during the First World War Era

Ulf Schmidt
JUSTICE AT NUREMBERG
Leo Alexander and the Nazi Doctors' Trial

Steve Tsang *(editor)*
PEACE AND SECURITY ACROSS THE TAIWAN STRAIT

C. W. Braddick
JAPAN AND THE SINO-SOVIET ALLIANCE, 1950–1964
In the Shadow of the Monolith

Isao Miyaoka
LEGITIMACY IN INTERNATIONAL SOCIETY
Japan's Reaction to Global Wildlife Preservation

St Antony's Series
Series Standing Order ISBN 978-0-333-71109-5
(*outside North America only*)

You can receive future titles in this series as they are published by placing a standing order. Please contact your bookseller or, in case of difficulty, write to us at the address below with your name and address, the title of the series and the ISBN quoted above.

Customer Services Department, Macmillan Distribution Ltd, Houndmills, Basingstoke, Hampshire RG21 6XS, England

Moving the Maasai

A Colonial Misadventure

Lotte Hughes

In association with
Palgrave Macmillan

Softcover reprint of the hardcover 1st edition 2006 978-1-4039-9661-9

First published in 2006 by
PALGRAVE MACMILLAN
Houndmills, Basingstoke, Hampshire RG21 6XS and
175 Fifth Avenue, New York, N.Y. 10010
Companies and representatives throughout the world.

PALGRAVE MACMILLAN is the global academic imprint of the Palgrave Macmillan division of St. Martin's Press, LLC and of Palgrave Macmillan Ltd. Macmillan® is a registered trademark in the United States, United Kingdom and other countries. Palgrave is a registered trademark in the European Union and other countries.

ISBN 978-1-349-54548-3   ISBN 978-0-230-24663-8 (eBook)
DOI 10.1057/9780230246638

This book is printed on paper suitable for recycling and made from fully managed and sustained forest sources.

A catalogue record for this book is available from the British Library.

Library of Congress Cataloging-in-Publication Data
Hughes, Lotte.
    Moving the Maasai : a colonial misadventure / Lotte Hughes.
        p. cm.—(St. Antony's series)
    Includes bibliographical references and index.

    1. Land tenure – Kenya. 2. Masai (African people) – Land tenure.
3. Masai (African people) – Claims. 4. Kenya – Colonization – History.
5. Great Britain – Colonies – History. I. Title. II. Series.

HD983.Z63H84 2006
967.62'004965—dc22   2005054653

10 9 8 7 6 5 4 3 2 1
15 14 13 12 11 10 09 08 07 06

I dedicate this book to the Maasai, and to
Norman Leys

Contents

Part II   The Aftermath

Part III   Interpretations

Abbreviations and Style

Abbreviations

ADC	Assistant District Commissioner
A/R	Annual Report
ASAPS	Anti-Slavery and Aborigines' Protection Society
BEA	British East Africa (also called East Africa Protectorate)
BPP	bovine pleuro-pneumonia
CO	Colonial Office
DC	District Commissioner
EAP	East Africa Protectorate
EAS	East Africa Syndicate
ECF	East Coast fever
FO	Foreign Office
GEA	German East Africa
IBEAC	Imperial British East Africa Company
JAH	Journal of African History
KAR	King's African Rifles
KLC	Kenya Land Commission
KNA	Kenya National Archives
NLC	Native Labour Commission (1912–13)
O in C	Officer in Charge
OUP	Oxford University Press
PC	Provincial Commissioner
PRO	Public Record Office, Kew, London (now renamed The National Archives)
RHO	Rhodes House Library, Oxford
SI	Stock Inspector
VO	Veterinary Officer

Style

Maasai is the correct spelling, but Masai will be used when citing records from the colonial era. Kikuyu is used in preference to Gikuyu. In bibliographical references and notes, co-publishers will not usually be given.

Glossary of Maasai Words

This is a basic glossary of the words most commonly used in this book, and other relevant terms, which mostly follows the Frans Mol dictionary of the Maasai language, Maa.[1] At times Maasai and others beg to differ with him; when in doubt, I follow Maasai preferences, particularly with proper names.

When quoting from colonial or any other texts which spell Maa words and proper names incorrectly, the original spelling has been left unchanged, together with anglicised forms such as *laibon* (loosely translated as prophet). There may well be other incorrect spellings in my text, which have been impossible to check – Maa spelling is often contested, not least among the Maasai themselves.

The prefixes are *a* – infinitive; *e/em/en* – feminine (plural forms *i/im/in*); *o/ol* – masculine (plural forms *i/il*). The alphabetical order given here is according to the root of the word, not the prefix, except for placenames.

ol-aiguenani, pl. *il-aiguenak*   age-set spokesman.

ol-aji, pl. *il-ajijik* age-set. Consists of a right-hand (senior) and a left-hand (junior) circumcision group, which finally join together in one age-set under a new name, usually at the *ol-ngesher* meat-eating ceremony (though this varies from one section to another). There is an interval of about 15–20 years between each new age-set.

ol-ashumpai, pl. *il-ashumpa*   light-coloured person. Originally used to refer to the people of the East African coast, and later to Europeans and Asians.

ol-ayioni, pl. *il-ayiok*   boy.

Il-Dorobo   people without cattle. A derogatory term for Ogiek, Mukogodo and other forest-dwelling hunter-gatherers and their descendants, though Maasai can 'become' Dorobo when they lose their cattle, and vice versa. Some scholars regard Dorobo as more of an economic category than an ethnic group. Also spelled Il-Torobo/Torrobo (sing. Ol-Toroboni/Torroboni), Ndorobo, Andorobbo. British colonial records commonly spelled it Dorobo.

en-kiguena, pl. *in-kiguenat*   meeting, discussion, consultation.

ol-kipiei   lung, lung trouble; bovine pleuro-pneumonia (BPP).

en-kishu/in-kishu   cattle; also the Maasai as a people. The section Uas Nkishu (Uasin Gishu), now living mostly in Trans-Mara, means patchy or striped cattle.

en-kiyieu ceremony of sharing the brisket, denoting close friendship.

en-kop, pl. *in-kuapi* land, country, earth, ground.

en-kutoto, pl. *in-kutot* a locality, whose resources are shared by local people.

Entorror the Maasai's former northern territory before 1911. Often refers largely to Laikipia, but in its fullest sense includes the central and northern Rift Valley from Naivasha to Lake Baringo. From *a-rror*, to trip or fall down.

e-manyata, pl. *i-manyat* a warrior camp. Often wrongly used to refer to an ordinary Maasai village. Anglicised as *manyata*.

ol-milo, or-milo tick-borne 'circling disease' in cattle. In veterinary terms, bovine cerebral theileriosis (ECF), though some say it is anaplasmosis.

ol-mumai, pl. *il-muma* oath. My informants used it to refer to the blood-brotherhood, or blood oath, allegedly made between certain white settlers and Maasai leaders (see Chapter 6).

ol-murrani, pl. *il-murran* warrior, anglicised as *moran*.

e-murrano warriorhood.

e-mutai destruction, disaster. Commonly refers to the epidemics of the late nineteenth century in Maasailand. From *a-mut*, to finish off completely.

Ngatet the area of western Narok to which many Maasai from Laikipia were moved in 1911–13. Loosely used by their descendants to refer to the south in general, as distinct from Entorror.

ol-ngesher ceremony at which the two circumcision groups that make up an age-set are joined together under one name, and senior warriors graduate to junior elderhood.

ol-odua rinderpest, gall-bladder; literally that which is bitter. From *a-dua*, to be bitter.

ol-oiboni, pl. *il-oibonok* loosely translated as prophet, ritual expert; anglicised as *laibon*.

ol-oirobi cold, flu, foot and mouth disease.

Olorukoti (also *Olorukoti loo Siria*) The Trans-Mara area of south-west Kenya.

ol-osho, pl. *il-oshon* an autonomous socio-territorial section of the Maasai. Today there are around 22 sections (though some scholars list 16 and 19) in Kenya and Tanzania, of which the Purko is one of the largest.

ol-otuno (*ol-otuuno*) ritual leader of the age-set, chosen at the *eunoto* ceremony.

ol-payian, pl. *il-payiani* elder.

ol-piron, pl. *il-pironito* firestick, used to kindle a new fire. Refers also to the 'godfathers' of an age-group, two age-sets senior, who kindle the new fire for boys when their circumcision period opens. They then act as godfathers for the group and are responsible for guiding them as warriors.

ol-porror, pl. *il-porori*   circumcision groups. The senior (right-hand) and junior (left-hand) eventually combine to form one age-set.

ol-purkel   lowland, wet-season pasture.

e-sirit, pl. *i-sirito*   a company of warriors, anglicised as *sirit*.

o-supuko   highland, dry-season pasture.

ol-tikana (also *en-tikana*)   East Coast fever in cattle, malaria in humans.

ol-torroboni, pl. *il-torrobo*   fly, tsetse fly, mosquito; also trypanosomiasis. See Il-Dorobo.

e-unoto   one of the ceremonies that marks the transition of warriors to elderhood, by upgrading junior warriors to senior warriors. Literally means 'the planting' or establishing.

Acknowledgements

Many thanks to all my informants, without whom this would be a much poorer piece of work and to my father David Hughes for the maps.

I owe a special thank you to the family of Parsaloi Ole Gilisho and to friends in the Lemek, Mara and Narok areas. Also to my research assistants and translators, who were primarily David Ole Kenana, Charles Ole Nchoe, Dan Ole Njapit, Francis Ole Koros and Martin Ololoigero. Assistance was also given by Helen Kipetu, John Ole Kimiri, the late Saiguran Ole Senet, John Ole Sayiaton, Dickson Kaelo, Elizabeth Sialala, and more informally, James Ole Lemein, Irene Lankas and Esther Nchoe. Vincent Ole Ntekerei, John Ole Karia, Joseph Ole Karia, David Paswa, Partalala Ole Kamuaro and Samson Ole Mootian gave expert advice and invaluable comment.

For their hospitality and support, thanks go to the Nabaala family; Brendan Carden; Lord and Lady Delamere; 'Jock' and Enid Dawson; Debbie Nightingale; Paul Lane and all staff of the British Institute in Eastern Africa (BIEA), Nairobi. Further thanks go to John Lonsdale, Patrick McAuslan, Dorothy Hodgson, Mick Thompson, David Turton, Ted Milligan, Father Frans Mol, Desmond Bristow, Glyn Davies, Walter Plowright, Brian Perry, June Knowles, Christine Nicholls, Taiko Lemayian, Justice Ole Keiwua, Nathan Ole Lengisugi, Keriako Tobiko, Hugh Ross, Veronica Bellers, Alison Davies, Deborah Colvile, Jacqueline and Kuseyo Ole Sasai, Daniel Salau, Francis Sakuda, and other members of the Maa Development Organisation besides those already named.

I am indebted to my doctoral supervisor William Beinart for believing in the subject and its potential, and to my examiners, David Anderson and Megan Vaughan.

I am especially grateful to the Honourable Mrs E. A. Gascoigne for permission to reproduce material from the Harcourt Papers; to Mark Harvey for granting permission to quote from letters written to Edmund Harvey by Norman Leys; and to Colin Leys for allowing me to quote from Norman Leys's papers.

Finally, I appreciate the financial assistance provided by the British Academy, the Beit Fund, the BIEA, the Kirk-Greene Travel Fund, St Antony's College and the Welsh Writers Trust which, in giving me the John Morgan Writing Award 1996/7, funded my pre-doctoral fieldwork. Warmest thanks to you all, and many others who cannot be named individually.

Preface

> The implications of the outcome of the 1913 case are still as fresh to the Maasai as your last dinner. The Maasai now feel more capacitated than they were a century ago. They now have their own lawyers and other experts, and the learned few want to revisit the case for fair judgement. There is no room for rest until and when justice is dispensed.[1]

The story of how the Maasai of Kenya lost much of their land and a 1913 legal challenge came full circle in 2004. Maasai lawyers and activists planned to mark the hundredth anniversary of the first forced move of 1904, which signalled the start of major land alienation, by suing the British and Kenyan governments for recompense and, if possible, the return of any land that was still available. In the event, the year passed with much talk but no legal action; the only activities to cause a stir took the form of ranch invasions and public demonstrations. There was a flurry of angry memoranda and exchanges in the media between politicians, ranchers and activists. These developments were not foreseeable when I wrote the thesis upon which this book is based, but they make my subject all the more relevant to the present day. (For reasons of space, two chapters from my thesis have been omitted in the production of this book. They discussed the life of Parsaloi Ole Gilisho and uprisings post-1913; readers interested in these subjects should consult my thesis.) History is still unfolding, and both informing and driving twenty-first century politics.

The proposed lawsuit is likely to involve revisiting the judgement in the so-called Maasai Case of 1913 in the High Court of British East Africa (see Chapter 4). Although Britain is seen as the main culprit of the piece, the Kenyan government is also targeted because the Maasai community claims that post-colonial regimes – together with individual land-grabbers – have continued the process of land alienation that the British began. The plaintiffs intend to use my historical evidence in an attempt to prove that an injustice was perpetrated, since no one else has previously gathered so much material on the subject, both oral and archival. But I am not involved in the claim *per se*; what people choose to do with the information I have shared is up to them.

This grievance and the desire for justice have simmered for a century, largely unseen and unheard outside Maasailand. There have been intermittent public outbursts, notably in the 1930s when the Kenya Land Commission took evidence from Maasai, and in the early 1960s at talks in London on Kenya's independence when Maasai representatives raised many of the issues that have resurfaced now. Much earlier, certain administrators in British East Africa foretold what might happen if the Maasai were forcibly relieved of their best land. Frederick Jackson, then Deputy Commissioner of the Protectorate, warned in 1903 against 'committing a gross injustice on the Masai' by giving way to settler claims to the Rift Valley. He wrote the following year: 'The Masai will never give us serious trouble so long as we treat them fairly and do not deprive them of their best and favourite grazing grounds ....'[2] In 1910 Arthur Collyer, District Commissioner (DC) for the Maasai's Northern Reserve on Laikipia and a man with great sympathy for Maasai interests, advised the government that the community should not be moved for a second time; in his view they did not deserve to be so shoddily treated. In 1928 the Narok DC, Major Dawson, remarked that continuing Maasai antipathy to government stemmed from 'the unfortunate circumstances of their move from Laikipia and the litigation which succeeded it. These circumstances have been remembered by the old men and suspicion of everything emanating from the Government is imbued into the children with their mothers' milk. Only time, sympathy, and continuity of policy can break down this tradition.'[3]

This has not happened. On the contrary, one organic end-result of imperialism for millions of subjects was the growth of literacy, political consciousness and – in the case of indigenous peoples – a desire for self-determination, which has meant that their relationship with the state often continues to be problematical. The decolonisation process that followed World War II led to the creation of new nation states, in which indigenous communities have fought to assert their rights, separate identity and cultural heritage in the face of attempts by governments and mainstream society to assimilate, marginalise or even exterminate them. Rights to land and natural resources are central to this struggle. The second half of the twentieth century saw the rise of the global indigenous rights movement, which was in turn inspired by earlier social movements, and supported by international human rights groups and non-governmental organisations. New technology has made it possible for indigenous communities, once remote and isolated, to make common cause by contacting other groups, mobilising and campaigning via the internet and 24-hour news media. In some ways the movement is a product of globalisation.

Developments within the United Nations also underpin the growing international recognition of indigenous rights. The UN set up a Working

Group on Indigenous Populations in 1982, which began drawing up a Draft Declaration on the Rights of Indigenous Peoples three years later. This has not moved beyond draft form. But a milestone was reached in 2000 when the UN established a Permanent Forum on Indigenous Issues, giving aboriginal peoples official recognition on the world stage. The International Year of Indigenous Peoples was marked in 1993, and the first Decade of Indigenous Peoples celebrated from 1995–2004. The Maasai see their action as a fitting punctuation mark to that. (A second Decade has since begun.)

Of course, none of this is straightforward. The story of this particular dispossession never was. It is mired in myth and myth-making, contested constructions of history, nationalism and ethnic identity, arguments over the meaning of land ownership, 'rational' land use, stewardship of wildlife resources and different concepts of justice. It provides a fascinating case study in colonial misadventure and its long-term repercussions, which has implications that extend far beyond Maasailand.

Maps

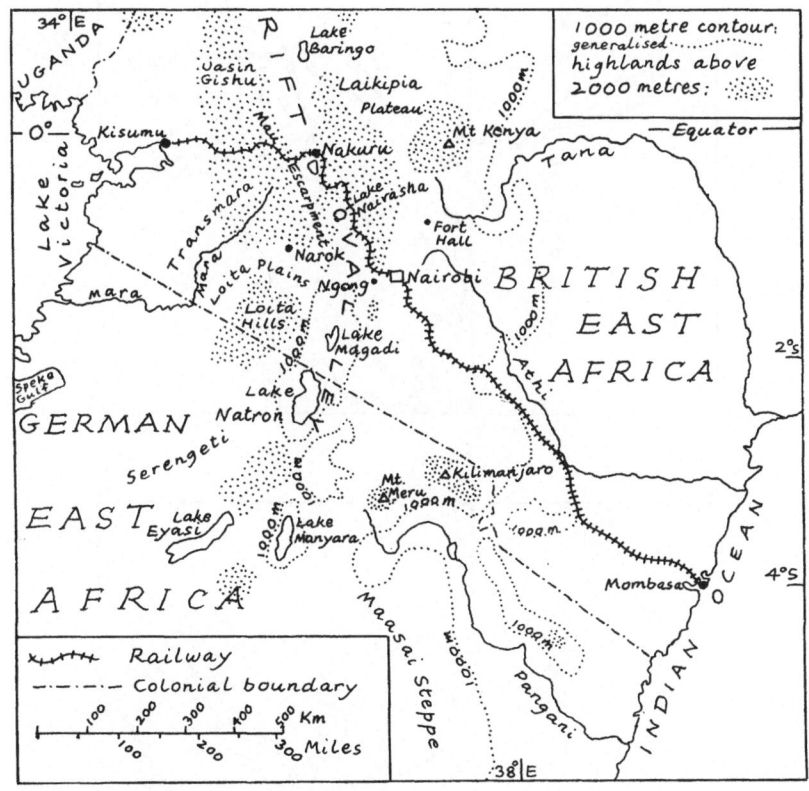

East Africa pre-World War I. Map drawn by David Hughes.

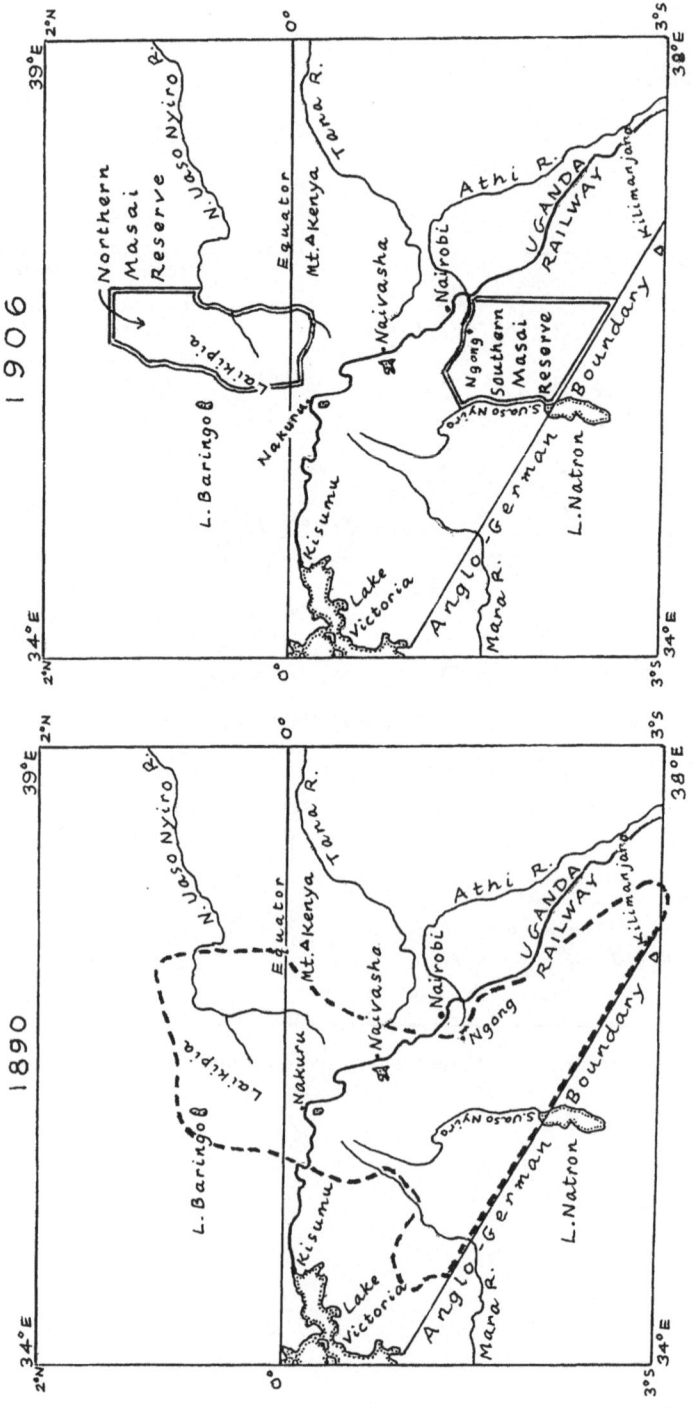

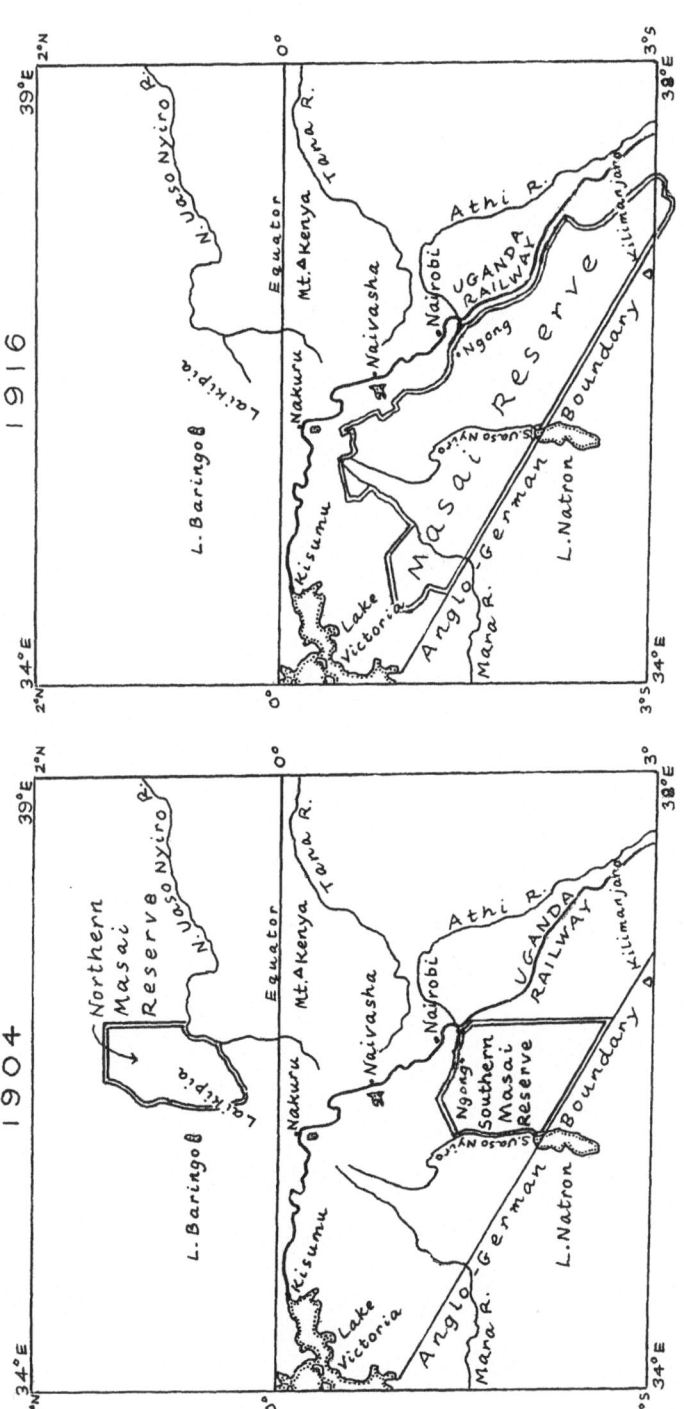

The distribution of the Maasai in British East Africa, showing how their territory changed over time. The broken line on the first map (1890) indicates the area in Maasai occupation before reserves were created. Masai was the colonial spelling, but Maasai is correct usage today. Maps drawn by David Hughes, after Sandford (1919).

Part I

The Moves and What Led Up to Them

1
Introduction

> I have no desire to protect Masaidom. It is a beastly, bloody system, founded on raiding and immorality, disastrous to both the Masai and their neighbours. The sooner it disappears and is unknown, except in books of anthropology, the better.
>
> Sir Charles Eliot to Lord Lansdowne,
> 19 April 1904[1]

Maverick colonial servant Dr Norman Leys once lamented that the true story of how the British relieved the Maasai of their land would never come out, despite his best efforts to publicise it. He wrote to his friend, the British MP and Quaker Edmund Harvey: 'Things aren't bad enough *yet* to give the chance of a scandal. Ten years more and somebody will write a sensational novel or there will be a native rising or in some other way the British public will get disillusioned.'[2]

This is not the novel he hoped for, but its contents are sensational none the less – and more shocking than fiction. It aims to pick up the investigation where Leys left off, frustrated by official obstruction and threats. Using his own unpublished evidence, Maasai and other oral testimony, and archival sources in Britain and Kenya, this study aims to produce new knowledge about the events that cost Leys his career and the Maasai the best part of their land. Though fragments are already known and documented, this book tells a previously untold story of 'white mischief' and the making of the so-called White Highlands, carved out of largely Maasai territory. Of course, some will argue that this was not originally Maasai land, that they were marauders who had stolen it from others, and that they cannot claim to be first comers to this territory. The short answer is that Britain recognised the East African highlands as Maasailand from at least the mid-nineteenth century, and this single word (or simply Masai) is writ large across early maps of the region. But no, the Maasai cannot claim to be first comers; possibly only forest-dwelling hunter-gatherer communities, collectively termed 'Il-Torobo', can claim that distinction.[3] Longer answers will become clear as the story unfolds.

The Maasai people have attracted enormous interest from travellers, traders, missionaries, administrators, historians and anthropologists since the earliest days of their contact with Europeans. Like Zulu warriors, they have almost come to represent Africa for those – largely in the Western world – who know little about the continent. Photographers, researchers, writers, fashion designers and tourists have followed those early adventurers, reproducing images and accounts that have in turn generated a welter of popular interest and representation.[4] In some of this material, exoticisation and demonisation of the Maasai (or who the Maasai are imagined to be) manifest as two sides of the same coin. All in all, the Maasai tend to remain fixed in time and space as archetypal noble savages, embedded in Western images of Africa, exploration and wilderness. Some Maasai themselves play up to this, both in order to satisfy tourist appetites and thereby capitalise on Western fantasies, and to invoke the idea of a 'traditional' idyll which has been shattered by modernisation. Public fascination focuses on whether they have 'moved on' at all since the turn of the last century, when Commissioner Charles Eliot dismissed their 'bloody system', or whether they still adhere to a supposedly timeless, traditional way of life. People are also curious to know whether the Maasai still exist, since news of their imminent extinction has been broadcast since the 1900s, and still abounds today. Some Maasai deliberately invoke this idea when calling for special protection as an indigenous community within the nation state. Little is popularly known or cared about their recent political history.

The ancestors of the Maasai (or more properly Maa-speakers) came to East Africa from southern Sudan sometime during the first millennium AD. They 'settled' in what is now Kenya and Tanzania, and continue to live there today, the great swathe of Maasailand broadly following the line of the Rift Valley and fanning out on either side. By the early nineteenth century, at the height of their power, they lived in and on either side of the Rift, occupying an area stretching from Lake Baringo in the north to central Tanzania in the south.[5] This former territory has been described as lying at a latitude of between one degree north of the equator and about six degrees south, and more than 200 km wide in some places.[6] Beyond mentioning some key events, I do not intend to cover their pre-twentieth-century history in any detail; this has been ably done by other scholars such as Berntsen, Waller, Galaty and Jacobs.[7] My interest is primarily in the early colonial history of the Maasai in British East Africa (BEA, later Kenya), particularly that of the Purko section, and their relationship with the British from the 1900s. Though some of this period has been thoroughly researched, there are major omissions in the historiography. There is little mention in the published literature of two of the most momentous events in the last hundred years of Maasai history: the forced moves which robbed the 'Kenyan' Maasai of the greater part of their territory, and resistance to the second move from Laikipia to the Southern Reserve which culminated in a 1913 court case brought by the Maasai, with the assistance of Leys and other Europeans in and outside the colonial service. The few historians who have covered these events fail to analyse fully their significance and effects, or to include a Maasai perspective and direct, attributable quotes. The resistance is typically dismissed as insignificant, largely assumed to end at the court case, and not placed within the context of other African resistance movements.[8] Sorrenson, for instance, writes: 'Their losses of land did not breed in the Masai [*sic*] a spirit of rebellion. Unlike the Kikuyu, they did not attempt to acquire European techniques to settle their grievances. The Masai reacted to the new society by ignoring it. Certainly the Masai court case was an exception but in this the Masai acted on the advice of Leys and Morrison, as a last desperate effort to retain Laikipia.'[9] It was in fact a major exception, which disproves his previous claim. By emphasising the roles of European supporters, Sorrenson downplays African agency. Many gaps remain which need to be filled.

Outlining the story

Briefly, the facts are these. In 1904–05, the British forcibly moved certain sections of the Maasai out of their favourite grazing grounds in the central Rift Valley (Naivasha–Nakuru) into two reserves in order to make way for white settlement. One reserve was on Laikipia in the north, the other in the south on the border with German East Africa where other Maasai sections already lived. Under a 1904 Maasai Agreement or treaty, these territories were promised to the people for 'so long as the Masai as a race shall exist'. Seven years later, the British went back on their word and moved the 'northern' Maasai again, at gunpoint, from Laikipia to an extended Southern Maasai Reserve. The second move was not completed until 26 March 1913. White settlement of the highlands was the primary reason for the expulsion; other reasons will be discussed later.

The second move was sanctioned by a 1911 Agreement, which the Maasai later claimed their leaders signed under duress. This Agreement effectively rendered the first one void.

As a result of these two moves, and later forced moves of communities including the Uas Nkishu Maasai from a reserve at Eldama Ravine to Trans-Mara, and the Momonyot of the Loldaika Hills to the same area, the Maasai of BEA lost at least 50 per cent of the land they had once utilised. (I have not investigated these and other moves of affiliated groups; they will only be referred to where relevant.) Some might inflate this estimate to nearer 70 per cent. It is difficult to come up with an exact figure since land in Maasai use, as opposed to occupation, before 1904 was never surveyed and officially quantified. Today, Maasai point to the fact that many Kenyan place names, including Nairobi, are derived from the Maa language as proof that certain lands were once theirs. Maasai leaders made this point much earlier, in their 1932 memorandum to the Kenya Land Commission (KLC).[10]

The British expected the Maasai to resist violently, as befitted their bloodthirsty reputation. This had partly been created by early coastal traders, and amplified in racy, best-selling accounts by nineteenth-century travellers such as the Scottish geologist Joseph Thomson.[11] Administrators were taken aback when a small group of young men hired Mombasa-based British lawyers and took the government to the High Court in 1913 to contest the legality of the second move and demand compensation for stock losses and depreciation in stock values as a result of the move. The plaintiffs tried to regain Laikipia, claiming that the 1911 Agreement was not binding on them and other 'northern' Maasai who had not signed it. They lost on a technicality, went to appeal, and lost again. However, this was a landmark legal action, apparently the first of its kind brought by indigenous people against colonial rulers in East Africa.[12] It was led by illiterate senior warriors, and initially launched by a charismatic age-set spokesman named Parsaloi Ole Gilisho. In an ironic twist of the tongue, the British anglicised his name as Legalishu. The full significance of the case has not been examined before, while Ole Gilisho is barely mentioned in histories of the Maasai or East Africa.

Illiteracy placed the Maasai at a major disadvantage in their battle of wits with the British, and contributed to their losing the court case. They could not write down their version of events, which has left the whole story largely in Western hands and archives to this day. They were forced, in their dealings both with government officials and their own lawyers, to depend upon translators and semi-literate mediators whose reliability was questionable. They could not record the numbers and names of people who allegedly perished on the Mau escarpment in the summer of 1911 during the second move, and so were unable to prove that anyone died at the hands of the British (if indeed anyone did, see Chapter 2). They attempted to enter into the colonial discourse through sophisticated verbal debate, as is evident from the allegedly verbatim accounts of meetings between Maasai leaders and British representatives in the run-up to the moves. Public debate and discourse are cornerstones of Maasai society and customary justice, and Maasai leaders assumed they could negotiate, and achieve justice, by employing similar tactics with Europeans. Certainly, there was plenty of talk. But literacy, and the ability to reflect upon and disseminate a contested text, enables a hegemonic power to out-manoeuvre its illiterate opponent more often than not. Things were about to change, but not fast enough in this case. As Leys said of the growth of literacy: 'A new air is blowing in the world. Any people who, like the Africans of Kenya, are determined to learn to read will soon be determined to be free.'[13] Maasai desire for self-determination preceded their literacy, but fell at this first hurdle.

Approach and timescale

My main focus is the forced moves, particularly the second; resistance culminating in the 1913 court case; and the repercussions of these events and associated land losses. But important subsidiary themes emerged in the course of investigation. These included allegations that Maasai leaders and certain white settlers became blood-brothers in an oathing ceremony held sometime before 1911 (see Chapter 6). This came out of oral testimony, and is barely mentioned in any written text. Making blood-brotherhood was practically *de rigueur* among early European explorers, professional hunters and administrators in the region, as they sought to make treaties with 'tribes' and establish the British flag – or in the case of hunters, simply forge useful friendships in dangerous territory. But its significance was completely different in the Maasai context as European settlers sought acceptance from their African neighbours, who in turn sought to make peace with these strange, and potentially threatening, immigrants. It seems the Maasai mistook settlers for government officers and believed that they were, in making blood-brotherhood, forging an official contract. Some elders told me they believed the blood-brotherhood bond still existed, and evidently set more store by this than the official Agreements. If this ceremony happened, it partly explains why the Maasai did not violently resist

European intervention. Even if it did not, stories about it signify an important social metaphor.

Other crucial subsidiary themes also jostle for attention. They include the environmental and disease impacts of the forced migration, not only for the migrants but also for other Maasai groups into whose territory they were pushed; the role played by European and African dissidents in challenging colonial policy in this period, and the ways in which individual whistle-blowers connected to the broader anti-imperial movement; the link between land grievances and the battle between British forces and Maasai warriors at Ololulunga, western Narok, in 1918; and the return of significant numbers of Maasai to their former northern pastures after World War I, where many worked for European farmers and went on to re-establish a community. Throughout, I am as interested in exploring people's perceptions of events as in finding out what actually happened.

In order to present the main and subsidiary themes as clearly as possible, the book is divided into three parts. Part I will investigate the moves and what led up to them. Part II will examine the aftermath of the moves, including the court case and environmental impacts. Part III will look at various interpretations, including examining what blood-brotherhood represented, and trace the trajectories of Maasai who reversed the exodus from Laikipia to return north. The story is told largely by interweaving and comparing these contested narratives (that is, those of the Maasai, British administrators, dissidents, settlers, scientists and others). There was also contestation within all these categories; for example, there was little or no consensus on administrative policy within British government circles either in BEA or in London. The Maasai also disagreed among themselves about the Agreements, the wisdom of the second move in particular, which leader to follow, how to respond to European intervention, and whether or not to resist. These disagreements continue today, when many Maasai are unable to discuss these events (now popularly linked to long-term underdevelopment, impoverishment and marginalisation) without blaming somebody. Dissidents – both African and European – had different ideas about methods of resistance. Leys in particular constantly wrestled with the dilemma of what he should do, with whom, and how.

My immediate focus is 1904 to 1918, with the broader frame of reference the last quarter of the nineteenth century to the present day. The story largely features events and personalities in East Africa, and I shall not dwell upon the British political scene, or individuals and organisations in Britain, in this period. The style of language is deliberately less academic than the doctoral dissertation upon which this book is based. I make no apology for sounding 'journalistic'; I am also a journalist, and wish to make the story both readable and widely accessible.[14]

Sources: oral

My informants included Maasai elders old enough to have taken part in the second move as small children; many have since died. Interviews were also conducted with members of Ole Gilisho's family (primarily his sons), and with the descendants of Ole Gilisho's son-in-law Ole Nchoko, who became first plaintiff in the 1913 case. Their testimony is rich, and touches on many other interrelated subjects. It raises issues that do not appear, or are barely mentioned, in any written text. There are almost no Maasai voices in the existing literature, which weakens an already sparse resource. This deficit also applies to the wider literature; that is, the historiography of the Maasai as a whole, quite apart from accounts of the moves and court case. In this regard, an example to scholars is set by the non-academic writer Gerald Hanley, who includes many direct quotes from Maasai and the full transcript of his interview with Marianyie Ole Kirtela, an interpreter for the British at Rumuruti and (according to King) an advisor to Ole Gilisho.

Scholarly exceptions include Brockington's work on Mkomazi, Tanzania.[15] In a key article, Berntsen lists and names his informants, and states where the full transcripts are available to other scholars.[16] Tignor also names his informants, but does not use direct quotes.[17] King tends to name his informants, and uses some direct quotes. Jacobs quotes his informants in composite form, but does not attribute information to named individuals other than Justin Lemenye, administrator Hollis's chief informant. There may well be other exceptions. Overall, however, as a result of the widespread omission of directly quoted testimony, I cannot do what Jan Vansina exhorts scholars of oral tradition to do: examine 'the whole corpus, or at least a large corpus of recorded tradition' in order to understand the testimony one has gathered in relation to 'all others'.[18] The term 'recorded tradition' is also problematic; in this literature it tends to refer only to stories about nineteenth-century and earlier history. There is an abrupt cut-off at 1900.[19]

The paucity of Maasai voices results in another major omission: little or no sense of Maasai conceptualisation of their colonial experience in general, and of the moves and resistance in particular. Their version of the story has largely been expunged from the historical record. By listening to and citing oral testimony, one can begin to excavate Maasai conceptualisation and move beyond an appraisal of the material facts (such as who moved, how many stock died, what diseases there were in those days, who said what to whom) to a perceptual realm. Material 'facts' dominate the literature, rendering many texts curiously one-dimensional in their pursuit of history as a reconstruction of events. This book attempts to redress the balance. While covering a sequence of events from several perspectives, it aims to add a metaphysical and perceptual dimension to the literature, and allow space for people's perceptions in the belief that these form a major part of 'reality'.

Within the performance that accompanies the delivery of oral history ('history cannot be divorced from the circumstances of its telling'), there was a definite sense in my interviews of very elderly Maasai wishing to have the last laugh before they faced death.[20] They did so in part by calling Europeans fools. With both African and European informants, there was also a knowing theatricality in the interview encounter.[21] The great age of interviewees (many in their 80s and 90s, some over 100) added poignancy to the performance and was possibly an incentive to share information; some elders were literally on their deathbeds. Therefore, people may have been keen to unburden themselves in the knowledge that I and my Maasai assistants intended to disseminate what they had to say, and add value to it by placing it in the public domain. Moreover, by imparting information, interviewees were symbolically asserting their power and authority over the colonial discourse and transcript. As Tonkin puts it, 'the act of authoring is a claim to authority'.[22] Some testimonies could be categorised as hidden transcripts, a concept developed by James Scott to describe both speech and gestures/activities which are used largely by subordinate groups to counterpoint public transcripts created and controlled by the dominant class.[23] Of course, one must add a dose of scepticism: were informants telling the 'truth', or having a laugh at my expense, too? There is no knowing, except that a fairly consistent narrative emerged from information gathered from people who were mostly interviewed separately, often many miles apart, and (outside the Lemek area) often with no prior notification.

Some obvious limitations should be mentioned. My Maasai informants were largely members of the Purko section, which bore the brunt of the second move, together with a sprinkling of interviewees from other sections. They were also overwhelmingly male. Though women have stories to tell about how the moves affected them, and some are included here, they were not – so far as one can tell – party to the political discussions that surrounded the moves and court case and did not, at this time, play a key role in the public political arena. Since this is my primary focus, I make no apology for the gender imbalance that has resulted; this is simply a rather male narrative, and gathering more women's stories would be another story and a different task. Facing time, budgetary and other practical constraints familiar to any researcher, and lacking a good female assistant/interpreter, it made sense to narrow my enquiries. I do not purport to show a gendered, pan-Maasai picture, which is beyond the scope of this study. No scholar has yet produced a broader sectional overview that might tell a very different story of these events from a range of Maa-speakers' perspectives, and illuminate from other angles such issues as the internal decay and tensions within the Purko community in this period.

Maasai oral testimony is augmented by interviews with the descendants of leading settlers, including the current Lord Delamere, their former employees, and other Europeans such as retired veterinary officers and the great-nephew of Andrew Dick, reputedly the first white man to be killed by the 'Kenyan' Maasai, after the Kedong massacre of 1895. A freelance trader and former accountant with the IBEAC, Dick disobeyed British orders in attempting to take revenge for the massacre, and reportedly killed up to 100 Maasai warriors before being speared himself.

Sources: written

My other key primary source is a collection of letters written by Norman Leys to British MPs Edmund Harvey and Ramsay MacDonald, which were discovered in a family archive where they had lain unseen since they were written in 1910–14. I shall call them the Harvey Letters, since they were primarily addressed to Harvey. In this correspondence, which was intended to inform parliamentary debate in Britain, Leys championed the Maasai cause in particular, challenged colonial policy towards the Maasai, and heavily criticised other aspects of 'native' administration. They are highly revealing of his motivation and philosophy, and of his clandestine actions to defend African rights through a network of contacts. They augment what is already known about his activities in this period from his letters to Gilbert Murray (to whom a few of the Harvey Letters were copied), J. H. Oldham, the Anti-Slavery Society and others. They also provide answers to some key questions raised by scholars who have studied Leys, and fill crucial gaps in their work.[24] Leys introduced Ole Gilisho to a lawyer friend, Alexander Morrison, after suggesting to Morrison that the Maasai needed legal help. It was officially assumed that Leys had instigated the legal action, and this led to his effective dismissal from service in East Africa by transfer to Nyasaland in 1913. He was not actually sacked, but it amounted to the same thing; his career in the colonial service never recovered, much to his regret.[25]

From the Harvey Letters it is clear there was a great deal that the government did not know about Leys's anti-establishment activities. In his published writings, and in other private correspondence, Leys never fully revealed his role in what may be called the Maasai affair. Hence John Cell could write after editing Leys's letters to Oldham: 'My conclusion is that Leys did precisely what he admitted, no more' – that is, he had simply done the Maasai a favour by putting them in touch with a lawyer.[26] The Harvey Letters supply some of the missing parts of the puzzle. Most importantly, they retrieve this episode from the flames: in an apparent fit of despair towards the end of his life, Leys burned his large correspondence on African and imperial matters. The few letters that remain are all the more valuable for that.

My main secondary sources for information on the moves, case and administrative context are Sandford, Cashmore (unpublished), Mungeam, Sorrenson and, to a lesser extent, Tignor.[27] The few available Maasai texts may be written from a Maasai viewpoint, but are not necessarily more accurate as history. Civil servant George Sandford's contemporary account, an official history of the Maasai Reserve and its colonial administration based upon government papers and records, should be differentiated from those of professional historians. If Leys and his friend William McGregor Ross may be counted as historians, they also covered these events in some detail, in a highly partisan fashion.[28] Colonial archives (including government registers, official correspondence, White Papers, district and provincial annual reports and veterinary reports) are an extremely rich source of information, some of it surprisingly candid. Likewise, references to these issues in the memoirs, letters and diaries of early administrators and other players are sometimes full and frank.

Sandford was Private Secretary to the Governor when he wrote his so-called 'Blue book'. It was compiled from official papers, so one assumes that the views expressed were not his own, merely a repeat of what his superiors had written. A comparison of this authorised version of events with alternatives, both written and oral, reveals an astonishing tale of administrative bungling, lies and cover-up. The elements of political spin will become apparent later, but Sandford should be credited with some frankness. For example, he admitted government failure to adhere to the terms of the 1904 Agreement, in not keeping a connecting road open between the two reserves. He also admitted that the British, to their cost, wrongly recognised the prophet Olonana and later his son Seggi as Paramount Chiefs, because certain officers 'had not carefully investigated the facts of the case'. They had mistaken the authority of the prophets for political power, and backed the wrong men in making some of them chiefs – notably these two. Over time, it was realised that the power of prophets was largely magical and ritualistic, and not binding on the Maasai as a whole. Furthermore, Sandford (lifting Eliot word for word) noted that 'the centre of political gravity was not with the elders or chiefs but with a republic of young men' – the warriors – 'governed by ideas of military comradeship and desirous only of military glory'. By 1918, the British were forced to recognise the warriors because the elders were so weak.[29] It is clear from this and other contemporary accounts that the British did not know who they were dealing with, or understand the relative authority and representativeness of different leaders (prophets, elders, age-set spokesmen and other principal warriors). If some did know, they did not act upon this knowledge.

Maasai social system

To give a brief overview, the Maasai are a Nilo-Hamitic people, divided into socio-territorial sections or *il-oshon* that straddle Kenya and Tanzania. No one seems to be agreed on how many sections there are; the number given varies from 14 to 22, though some earlier sections are now defunct as a result of internal warfare and incorporation.[30] Each section enjoys grazing and other resource rights in a particular area, with sections subdivided further into localities or neighbourhoods called *inkutot*. As transhumant pastoralists the Maasai use land seasonally, moving from highland to lowland pastures according to the rains, which allows grazing and other vital resources to regenerate. The community is divided into five clans (there is some disagreement over numbers) and two moieties, red and black. The prophets of Mbatiany's family, including sons Senteu and Olonana, all belong to the *en-kidong* (pl. *in-kidongi*), a sub-clan of the Il-Aiser (anglicised as Laiser).

Maasai society and economy revolve around livestock, with cattle valued particularly highly as a mobile form of wealth, medium of exchange and marriage, source of food, symbol of relationships, and for their sacred significance. However, increasing numbers of people no longer follow an exclusively pastoral mode of life or restrict their diet to livestock products, if indeed they ever did. The first incomers to the region were originally agro-pastoralists, and pastoral specialism only developed later. Earlier ethnographers (including Jacobs, writing in the mid-1960s), tended to paint the Maasai as purely pastoral and contrast them with mixed economy 'Iloikop' or 'Kwavi' Maa-speakers, but this sharp division is no longer accepted as true. Also, the importance of their trading exchanges with neighbouring societies tended to be overlooked in the past, and their dietary preferences for blood and milk overstated. Like many pastoralists in Africa, the Maasai are 'exchange pastoralists' who culturally value milk and meat above all other foods but who have, for many years, varied their diet with foods exchanged for other goods when the need arose. Livestock is owned individually, the family being the principal stockholding group, but land was not 'traditionally' owned by any one person. Before individual land ownership was introduced, land was (and is still ideally) viewed as a community resource.

Customarily, the Maasai are acephalous and do not have 'chiefs' or headmen. These were only introduced by colonial governments; since independence, Kenya has perpetuated this system. Political authority 'traditionally' lay with councils of elders and age-set spokesmen, elected for their leadership qualities, while prophets wielded spiritual authority. The age-set structure is the fundamental organising principle of Maasai society, and instils values of egalitarianism, sharing and respect. Reference to age-set chronology dating back to the mid-eighteenth century is one of the only ways to ascertain what year an event took place, since elderly Maasai do not tend to think in terms of calendar years.[31] Women and girls do not belong to age-sets, although they pass through rites of passage that parallel those of males as they graduate from boyhood to elderhood, and join their husband's age-set on marriage. Councils of elders constitute the main decision-making bodies, though some of my evidence challenges this model and suggests that younger men also are or were central to decision-making processes. A few women are becoming more influential politically. The idea that Maasai and other pastoral societies in Africa are intrinsically patriarchal has been increasingly challenged by women anthropologists such as Hodgson, revisionists of an ethnography previously dominated by male scholars who tended to focus upon male roles – partly because their informants were predominantly male, too.[32]

The fluidity of ethnicity in Africa is implicit throughout this book. Maa-speaking peoples are characterised by fluidity rather than by fixed, ahistorical models, as other scholars have demonstrated.[33] But when describing informants' origins, I shall use their preferred terms; for example, the Il-Laikipiak (more properly Il-Aikipiak) is an extinct section said to have been wiped out by a combined force of Purko-Kisongo warriors in the mid-1870s, yet many individuals continue to claim that they are Ol-Aikipiani. There is growing Laikipiak nationalist sentiment in this community on Laikipia today. The non-governmental organisation Osiligi (Organization for Survival of Il-Laikipiak Indigenous Maasai Group Initiatives), one of the groups involved in making land-reparations claims against Britain in 2004, was created expressly in order to champion Laikipiak interests.

The place of Maasai in the early 'colonial' state

Until 1920, when BEA became a colony, the Protectorate was technically a foreign territory. It grew out of commercial interests pursued by the ill-fated Imperial British East Africa Company (IBEAC), founded in 1888 by Scottish shipowner William Mackinnon. Several 'Company men' (including Hobley, Ainsworth and Jackson) went on to become administrators in BEA and the neighbouring Uganda Protectorate. The establishment of BEA in 1895, and subsequent white settlement, was very much an afterthought to strategic considerations including the need to protect the Suez route to India, and repel largely French and German encroachment in this region.[34]

The fact that BEA soon became 'settler country' set it on a collision course with the Maasai. They simply did not fit into the new equation – though as Chapter 8 will show, many went on to reinsert themselves as labourers on European farms. As Berman and Lonsdale have written, a settler 'playground [was] carved by the colonial state from the dry-season pasturage of the former lords of East Africa, the Maasai'.[35] Within a short space of time white pastoralists had supplanted black ones in the country's most productive, resource-rich heartland of the Rift Valley and highlands, and private land tenure has taken precedence over communal land regimes from that day to this. Furthermore, up until April 1902 thousands of Maasai had lived in the eastern province of the Uganda Protectorate. Then the boundary changed to bring this province – which included most of the highlands, and the much-prized grazing grounds around Naivasha – into BEA, so that the new Uganda Railway (see next chapter) could be managed by one administration. The railway ran through the best Maasai grazing grounds in the Rift, and divided the sections. Nothing was to be the same again. One wonders what might have happened if some of the Maasai sections had stayed on the Uganda side of the border, where European settlers never held sway. Uganda 'was considered unsuitable for European settlement' and was soon regarded as 'a black man's country to be developed along the lines of the West Coast territories'.[36] A different kind of state might have valued African pastoralism more highly, and supported the development of the sector – so long as it did not compete with European stock production, as it was seen to do in BEA.

One of the peculiarities of BEA was the relentless tug-of-war between powerful settlers, the local administration and the metropole. The settlers sought self-rule, their unofficial leader Lord Delamere telling his wife Glady years later: 'Of course if once we got any real control of any part they would never be able to stop us governing the whole. Don't mention this to anyone.'[37] The Maasai became pawns in that struggle, as Chapter 2 will explain, largely because of where they were on the map and how their lifestyle, cattle economy and economic contribution to the state were perceived. Race, or more precisely racial perceptions, were central to these early manoeuvres and the debates around them – though some officials initially advocated Indian rather than European settlement, hence this was not entirely a black and white issue. As for class, the Maasai did not fit the overlord–peasant model that characterised relations between colonial capitalism and Africans, simply because they were not poor peasant cultivators. Their wealth in cattle allowed them to remain relatively aloof from the state, until the single Southern Reserve was created, taxation increased, and freedoms were curbed by regulations governing labour, mobility, cattle trading and the rest.

The earliest relationship between the Maasai and the British had been relatively friendly, and featured a military and patron–client alliance. When British administration was first established, the Maasai were recovering from a series of devastating blights – rinderpest, bovine pleuro-pneumonia, smallpox and drought – following in the wake of civil warfare, which had brought them to their knees. These disasters were known collectively as *emutai*. Whole communities sought refuge in British forts and with neighbouring peoples, while many child 'debt pawns' were taken in by missions. Some prophets forged mutually beneficial alliances with white administrators; notably Olonana was made Paramount Chief (a nonsensical position, as it turned out), placed on the government payroll, and gained British backing in his protracted fight against his half-brother and rival, Senteu. Warriors were only too glad to lend their services to British punitive expeditions, which hired hundreds of them as auxiliaries, paid in raided stock. This enabled certain sections to rebuild their herds and strength, and some (notably the Purko) recovered remarkably quickly. Ole Gilisho himself led some of these mercenaries into battle on behalf of the British. Maasai were also hired and prized as interpreters, caravan guides, herders and personal servants. Colonial conquest had its advantages, then, but it put an end to Maasai domination of a space whose epicentre was the central northern Rift Valley.

These early alliances turned sour as settlers rode in on the back of the railway, and demanded their share of the Rift. The Maasai were assumed to represent a threat to the railway, lines of communication and European settlement around 'the iron road'. Also the vast cost of building the railway had to be recouped somehow. Official priorities shifted away from the protection of 'natives' to the promotion of commercial agriculture by Europeans on either side of the line. Eliot summed up this new thrust in 1903:

> East Africa is not an ordinary Colony. It is practically an estate belonging to His Majesty's Government, on which an enormous outlay has been made, and which ought to repay that outlay.[38]

Berman and Lonsdale's remarks on the early situation cannot be bettered. The state was 'institutionally incoherent', and brimful of contradictions. These included 'the contradictions between settler accumulation, which required great pressure on the supply of African labour, and the conditions for stable political control over African societies'. White settlement was believed to be an economic necessity, and meant 'the baronial consolidation of conquest', yet simultaneously it represented 'the chief threat to the politics of control'. It 'marked and maintained boundaries, the very essence of state-building' partly because whites created buffer zones between warring ethnic groups. Most crucially for pastoralists, 'white settlement would pin down pastoralism, the way of life that kept Africans idle, unnervingly on the move, and impervious to the benefits and constraints of civilization. The politics of conquest was brought symbolically to an end with the Maasai moves of 1904 and 1911. These fenced pastoralism out of the best grazing in the Rift while fencing capitalist ranching in.' As we shall see, however, in the long run 'pastoralism was not pinned down, it merely became subversive'.[39]

Critics of empire: Leys and McGregor Ross

These two men apparently first met in BEA in 1907, when Ross caught a fever and became Leys's patient at Mombasa.[40] Leys originally planned to collaborate with Ross in writing a book on Kenya together. He became frustrated and vexed when Ross did not come up with the goods quickly enough, as is evident in Leys's letters to Ross's wife Isabel, and in the end they published separately, three years apart.[41] Though critics of colonial policy in Britain were dismissed as sentimentalists who knew nothing of local conditions, the same charge could not be levelled at them. They, too, were 'men on the spot' whose experience lent weight to their opinion, as settlers were so fond of saying of themselves. And as colonial civil servants, they had access to information that was potentially explosive. They used this to great effect. Wylie credits them with '[helping] to thwart the acquisition of greater political power by settlers while African protest matured'.[42]

Leys was born in Liverpool in 1875, and as a child lived mostly with his grandfather, a Presbyterian minister in Lanarkshire, after his mother died giving birth to younger brother Kenneth. (Incidentally, Leys, McGregor Ross and Harvey were all close age-mates of Ole Gilisho: their life dates were respectively 1875–1944, 1877–1940, 1875–1955, and c.1875–1939.) There was a tussle between his grandfather and Scottish barrister father over custody of the two boys; when their father converted to Catholicism, the grandfather took them to America to prevent them from being reared as Catholics. Later, Leys studied medicine at Glasgow, where he first met Gilbert Murray (who became his life-long mentor and correspondent) and lived and worked in the slums – a formative experience.[43] He arrived in Africa in October 1901 to work as a doctor with the African Lakes Corporation at Chinde, Portuguese East Africa. He transferred to BEA in September 1905[44] and stayed for seven years as a government medical officer based successively in Mombasa, Nakuru and Fort Hall. He wrote an influential 1911 report on sanitation in Mombasa, which earned him government praise. In this and other reports he drew links between socio-economic conditions and ill health, which were by no means obvious to policy makers at the time.[45] After his own forced move to Nyasaland, he became interested in the Chilembwe uprising of 1915, interviewed many of the survivors in prison, and wrote about 'this new kind of unrest' as a footnote to his analysis of Kenya.[46] His later life, while a GP in Derbyshire, was devoted to writing and activism.

Leys devoted a chapter of his first book *Kenya* – which he had been planning since 1911 – to the story of the Maasai moves and court case and what led up to them. His major problem, as Wylie has pointed out, was the lack of official data to back up his claims about the effects of administrative policy upon the Maasai, or any other accurate facts and figures for Kenya. He had to rely upon estimates and hearsay, drew heavily on Sandford and complimented him: 'There are a few notable omissions in the story as told by the Blue book, but in the main it is candid and impartial as few official statements are.'[47] He repeated Sandford's account of the 1895 Kedong massacre (in which, after extreme provocation, Maasai warriors killed 456 men in a 1400-strong trading caravan of 'Swahili' and Kikuyu passing through their grazing grounds in the Kedong Valley) and the impression that subsequent British justice (the Maasai were exonerated) made upon Olonana.[48] Leys commented: 'Very typical is the immediate recognition by a savage people of a standard of justice higher than their own.' One can never be sure when Leys was being ironic. It seems unlikely here, despite his knowledge of the way in which the British legal system ultimately failed the Maasai, since he also declared: 'Railways and courts of justice are the two great boons our Governments have given to the people of tropical Africa.'[49] He believed the railways were a godsend because they saved the lives of African porters, and stopped the spread of disease along caravan routes.

Leys recognised Ole Gilisho as 'the most influential Masai in the northern reserve'.[50] His writings were also highly revealing of Arthur Collyer's role and attitude. As District Commissioner of the Northern Reserve and a Maa speaker, Collyer was probably the administrator who was closest to the Maasai, and most genuinely concerned with their welfare. Leys described Collyer's disquiet over the moves, and suggested Collyer was his main informant in the administration. Although this admission came many years after Collyer's premature death from tuberculosis in September 1912, official knowledge of Collyer's relationship with Leys may well have cost him his job with the Maasai, and the accompanying distress may even have hastened his end, as Collyer's family believe today.[51] Collyer had promised Leys he would write the Maasai chapter for *Kenya*, but after his death it had to be written without his help. Leys quoted from a private letter Collyer sent him just before the final move, in which his disgust was palpable:

As regards the Masai move, this sudden change of front has staggered me though I hold it was the right thing to do, if done long ago. The manoeuvres, etc., that have been employed with regard to the Masai have sickened and embittered me. I have always said that the policy of putting the Masai into one area was right, but I cannot uphold the methods that have been employed to bring this about. If in five years' time you write a book and I am in a position to give you the information on the Masai, you shall certainly have it.[52]

Leys obscured his own role in fomenting opposition to the Maasai moves. But he rather gave the game away by issuing a warning to would-be protesters: 'Whether it was right or wrong to protest against the Masai move may be doubtful. In any case, the reader who may live and work in Africa should be warned that if he ever takes a similar step he will do harm as well as good. If he feels he must, then he should. But the fact is that there is very little use in trying to stop these things. What is needed is rather to appoint governors and others in authority who will not attempt them.'[53]

McGregor Ross was a year younger than Leys, also of Scottish stock, born in Southport, Lancashire, in 1876. Having trained as a civil engineer, he came to BEA in 1900 to work as an Assistant Engineer on the Uganda Railway. He was Engineer-in-Charge of the Nairobi water supply from 1903. By the following year, at the age of 28, he had risen to become Director of Public Works, but was forced to resign in 1923 after agitation from settlers over his alleged mismanagement of public funds. McGregor Ross's face had never fitted the colonial scene. He was fiercely teetotal, moralistic, bookish, aloof from clubhouse camaraderie, and seen as a bit of a prig. (The evidence for this is in Wylie, but more directly in Ross's diaries of his early years in Africa.) As a member of the Legislative Council from 1916–22, a position that came with the job, his politics were seen to be soft on 'natives'. It was here that he became more and more anti-settler, forced to listen to 'the crude, crass clamour of self-interest'. In a progression that paralleled Leys' own political and philosophical maturation, his East African experience turned McGregor Ross into something of a radical and activist; he also became a Quaker and pacifist.[54]

In *Kenya from Within*, McGregor Ross devoted a chapter to the Maasai, focused on the moves, court case and their experience in the Southern Reserve. It was more broadly a diatribe against the arrogance of settlers and their abuse of privilege. Ross aimed to alert the British public to the fact that this class of mostly unelected men was running things in Kenya, to the detriment of Africans. His writing style was like a gossip-laced conversation, sarcastic and ironic by turns, punctuated by many exclamation marks and asides in brackets. It would never have worked to try and combine his style with that of Leys; maybe Ross knew this when he baulked at collaboration. His obituary writer noted: 'Pungent as were his criticisms, the book was marked by a racy humour that saved it from the bitterness to which the enthusiastic crusader may too easily become subject.'[55] This sounds like a dig at Leys.

McGregor Ross was clearly reliant on Sandford for much of his information, but he also made some fresh, key points. After Kedong, he claims that a naïve belief in the trustworthiness of the British government 'clung to the Masai for years' in the face of contradictory evidence. Speaking from experience of building the railway through Maasai country, he said they 'behaved in exemplary fashion, giving Government no trouble whatever'. It is important to record this peaceful response, since fear of Maasai violence towards white settlers around the railway, and their perceived threat to the line itself, was used to justify their removal from the Rift. On retiring, he threw himself into humanitarian and political activism in Britain and Geneva. He and Leys were members of, among other organisations, the League of Nations Union Mandates Committee and the Labour Party's Advisory Committee on Imperial Questions. Towards the end of his life, Ross became more concerned with peace activism, and spent less time advocating for Kenya's Africans.[56]

Parsaloi Ole Gilisho

Born Laikipiak but forcibly assimilated as a child into the Purko section, Ole Gilisho (c.1875–1939) was a member of the Il-Mirisho or right-hand circumcision group of the Il-Tuati II age-set, and an important age-set spokesman. He launched the initial legal action in 1912, but his son-in-law Ole Nchoko became first plaintiff in the case when it reached court, while Ole Gilisho became a defendant in favour of the action. He is still spoken of today as a folk hero and 'king' of the Maasai, who is celebrated in praise songs.[57]

There is little written information to suggest what kind of person Ole Gilisho was, never mind what motivated him. Therefore I have had to rely on oral testimony in order to flesh out his character and actions, while bearing in mind that my informants were likely to be biased towards him since they were largely Purko, and several were also of Laikipiak descent. In the early colonial record, apart from Frederick Jackson's glowing testimony, he was largely dismissed as a troublemaker, a conservative, and a lone voice in the wilderness who was not supported in his opposition to the British by the majority of 'northern' Maasai.[58] That attitude was to change in later life, when Ole Gilisho became something of a model elder in western Narok, though some local administrators continued to see him as difficult and obstructive.

From childhood, he appears to have been an unusual person. Father Frans Mol has written: 'Maasai informants say that his birth [on the Leroghi Plateau] was surrounded by a number of unusual circumstances. His mother is said to have been an *enkaibartani*, a young unmarried girl recently circumcised but still in the official status of partial recluse and untouchability after circumcision. Her circumcision may have been speeded up when she was found to be pregnant.'[59] This would actually make him unclean in Maasai eyes, since it breaks taboos around purity and should have precluded him from leadership in later life. But the Purko who snatched him during a raid may not have known the circumstances of his birth. His biological father was called Magiro (though one son, Mapelu, claimed he was Maatany), and his adoptive father was Leposo. His mother gave birth while crossing a stream or river. One informant said that he was pulled out of the water clutching stones in both hands, a mystical sign associated with prophets. He began exhibiting leadership qualities very soon after circumcision in 1896–97, was appointed *ol-aiguenani* or spokesman for the age-set, and kept the position for life. He went on to marry nine wives (some say 12), and had ten sons and eight daughters. I met his last surviving wife in 2000, when she was close to death. Senile, and unable to be interviewed, she lived with her son Leperes and his wife Kirapusho near Lemek, western Narok.

These three characters, especially Leys and Ole Gilisho, are central to the story that follows. Enigmatic, stubborn, often infuriating, supremely moralistic, and ahead of their time, they were to stir many passions and shake the very fabric of the colonial state.

2
The Moves

Meisho ilimot, inkulie ebaya
When events occur, only part of the truth is sent abroad; the
rest is kept back.

Maasai proverb[1]

Reports by nineteenth-century travellers and missionaries of the
environment and peoples of the highlands of East Africa were to heavily
influence the early administration of the Maasai, official and settler
views of them and their territory, and settlement itself. This chapter
touches on a key text before going on to describe how European settlement came about, the furious disagreements over land policy that forced
Charles Eliot out of office, the events that led up to the first Maasai
move, and why the second move went so disastrously wrong.

When the British arrived, they found the Maasai occupying land that
was ideal for white settlement – high, green and sweet, its climate a cool
relief from the humidity of the coast and a great deal healthier. Long
before settlement was considered, the highlands were being strongly
recommended to Europeans by early visitors. 'A more charming region
is probably not to be found in all Africa,' wrote Joseph Thomson after
his 1883–84 expedition for the Royal Geographical Society, which had
sent him to East Africa to find a direct route between the coast and Lake
Victoria. He is credited with being the first European to cross northern
Maasailand, although the German naturalist Dr Gustav Fischer had
taken the same route weeks earlier and been forced by the Maasai to turn
back at Lake Naivasha. As Thomson described it, the northern plateau
area including Laikipia had the look of a little Britain in Africa. His pen
dripped superlatives as he drew parallels between this landscape and
that of home (in his case, Scotland). It was a 'park-like country' complete

with 'flowering shrubs', 'noble forests', 'babbling brooks and streams' and 'pine-like woods [where] you can gather sprigs of heath, sweet-scented clover, anemone, and other familiar forms'. The familiarity of its flora contrasted with the strangeness of its people. They simply had to go.

Thomson wrote as if the local residents had already fled: 'The greater part of Lykipia – and that the richer portion – is quite uninhabited, owing, in a great degree, to the decimation of the Masai of that part, through their internecine wars ... .' Many Maa-speaking peoples had indeed perished in the Iloikop civil wars that ended in the 1870s. The Laikipiak section, which is identified primarily with the Laikipia and Leroghi plateaux, was virtually destroyed by a combined force of Purko-Kisongo Maasai, and the survivors assimilated by the Purko, Kisongo and other communities. The victors would have moved into the vacuum. Yet while reporting low population density, Thomson noted 'great herds of cattle, or flocks of sheep and goats are seen wandering knee-deep in the splendid pasture'.[2] There are no herds without herders, but unwittingly or otherwise Thomson seemed to overlook transhumant pastoralists' seasonal occupation and use of land. Lord Lugard did the same when writing glowingly of the Mau, an escarpment west of the Rift: 'This area is uninhabited and of great extent: it consequently offers unlimited room for the location of agricultural settlements or stock-rearing farms.'[3] Eliot declared: 'We have in East Africa the rare experience of dealing with a *tabula rasa*, an almost untouched and sparsely inhabited country, where we can do as we will ... .'[4] In this way another myth was propagated, which would prove fatal to the Maasai presence. Alfred Claud Hollis later acknowledged, and Sandford reiterated: 'The absence of fixed habitations and the periodical migrations of the Masai led to the belief that a considerable portion of Masai country was masterless. Consequently, Europeans began applying for areas that did not appear to be continuously occupied.'[5] While also mentioning that large parts of the highlands were empty, Harry Johnston – an eminent naturalist as well as an administrator – suggested that 'the celebrated Masai' should not be seen as a deterrent to Europeans fearful of their violent reputation: 'They want a little managing, that is all ... .'[6]

At first, despite what the early visitors thought, there was no question of ousting the Maasai to make way for white settlers. Initial concern focused on their nomadism, which rendered them and other wandering 'tribes' beyond government control. Administrator John Ainsworth suggested a remedy in 1899:

After a time when our Military forces are more organized and our Administration is more extended we shall be more able to edge in these nomad tribes and by degrees make it impossible for them to wander about without our permission, we could then clearly define the Masai-lands and see that the limits were kept. A policy of gradually bringing these people under our complete control is better than one of using absolute force at once ... .[7]

The century would turn before European settlement of upcountry BEA was seriously considered. Its primary value was strategic (see Chapter 1); it was a conduit between the coast and Uganda, source of the Nile. This was cemented by the Uganda Railway, completed in 1901. With its coming, the hub of government and commerce was to move from Mombasa to Nairobi, portal to the highlands, which had begun life as a shantytown for railway construction workers. Besides inviting Europeans to settle either side of the railway, Ainsworth encouraged Asians to set up shop at stations along the line, built largely with imported Indian labour. But Eliot advised that Indian settlement should be confined to the lowlands, and described the highlands as 'pre-eminently a white man's country' in his first annual report.[8] Eliot's policy of excluding Indians from the highlands was adopted by all his successors.

Land laws and early settlement

The Crown acquired, through legislation including the Indian Land Acquisition Act 1894 and an Order in Council 1898, all land in BEA apart from some coastal areas. Land taken for public purposes was said to be held in trust for the Queen. But the Crown did not yet have the power to alienate land; this was legalised by an Order in Council 1901. This defined Crown land as 'all public lands ... which for the time being are subject to the control of His Majesty by virtue of any Treaty, Convention, or Agreement, or of His Majesty's Protectorate', and all lands acquired by the Crown through the earlier acts. The meaning of the term public lands was not explained.

Under the Crown Lands Ordinance 1902, nearly 6000 square miles of land was alienated up to 1915. Settlers could obtain 99-year leases in lots of 640 acres for agricultural land and 5000 acres for stock farming, at rents varying from a halfpenny to tuppence an acre. Parcels of land up to 1000 acres could be bought outright for four shillings an acre. However, most of the early freeholds before 1912 were granted without payment, except for negligible survey fees. By 1905, when the Colonial Office took over the administration of BEA from the Foreign Office, government policy aimed to prevent land accumulation, stop dummying and revise rents when issuing new leases. In practice this was not carried out, and large concessions were given to individuals. In December that year, Lord Elgin (Secretary of State for the Colonies 1905–08) decreed that all grants or transfers of land which would allow one person to acquire more than 10,000 acres had to be approved by him. Dummying – the practice of applying for land in the names of relatives – was rampant. One result of this was a rise in absenteeism, and under-development of land, which was also against the rules. Many farms remained unoccupied. By 1906, Ainsworth was complaining that the best part of the Rift Valley was 'held up by absentees and ignorant squatters' and that the pastures had deteriorated since the Maasai had left. The Land Office tried to take action against absentees in the High Court, but many escaped conviction through loopholes in the law.

The Crown Lands Ordinance 1915 extended leases to 999 years, in response to settler demands either for freehold or long leases. Rents were set at tuppence halfpenny up to 1945, with subsequent variations fixed at a percentage of the unimproved value. Development conditions were virtually abandoned, though they had not been enforced at the best of times. For the first time, this law defined Crown lands as including all land occupied by or reserved for the 'native tribes'. Africans had effectively become tenants at will of the Crown. Much earlier, as it also attempted to do in British West Africa, the Crown had asserted its right to claim 'waste and unoccupied lands' as Crown lands, by virtue of its right to the Protectorate. The term 'unoccupied' was to cause untold grief for Africans who only seasonally utilised certain areas. The first Commissioner of BEA, Sir Arthur Hardinge (1895–1900), had declared that Africans only owned land so long as they occupied or cultivated it. The moment they moved off the land it became 'waste'.[9]

In reality, Leys later claimed, the Crown did not recognise *any* African rights of occupancy or ownership against those of the Crown, only against other Africans. He suggested that the only occasion on which the government had granted any right in land to Africans was when it made the 1904 Maasai Agreement, but that turned out to be illusory. The Crown was in fact the absolute owner of all land apart from those areas it had alienated. As for reserves, he felt the term was 'totally misleading; there was no other word for unalienated Crown land in native occupation'.[10] Several areas (including North and South Masai) were provisionally gazetted as 'native reserves' or Closed Districts under the Outlying Districts Ordinance 1902, but there was no legal provision for reserves as such until the 1915 Ordinance. In 1926 the government finally gazetted 24 native reserves covering 46,837 square miles, of which 14,600 were in the Southern Maasai Reserve.[11]

Before World War I there were various flirtations with proposals for Jewish, Punjabi, socialist 'Freelander' and Finnish settlement in the highlands, all of which came to nothing. Settlers furiously opposed plans for Indian and Zionist settlement in particular, and leading churchmen added their voices to the clamour. Zionists were offered the Uas Nkishu plateau as a homeland, but turned it down, much to Colonial Office relief. The earliest white settlers were predominantly South Africans, both Boer and British, who were joined by English and Anglo-Irish aristocrats and disparate 'roving adventurers, with little or no means' who had begun arriving in the 1890s.[12] The total numbers of European settlers in early BEA were negligible – by 1911, according to the census, there were only 428 'settlers, planters, farmers and gardeners' out of 3175 Europeans. The European population had risen to 5438 by March 1914, although only a fraction were settlers and their white employees. Alienated land was owned by a handful of people; by 1912, 20 per cent of it was held by just five individuals or syndicates.[13]

It was the coming of the railway that sealed the fate of the Maasai. It sliced their territory in two, made the highlands accessible and their settlement and economic development by Europeans possible. The Maasai prophet Mbatian (Mbatiany in Maa), father of Olonana, had foretold the arrival of the white man and the railway many years before. He had a vision of Europeans, represented by white birds, while the railway was seen as a great snake stretching from sea to lake.[14] His visions were indeed ominous. Eliot warned the Foreign Office in April 1904 that if the Maasai were allowed to keep the best land along the railway, while Europeans settled all around them, the latter were likely to launch 'a sort of Jameson raid' to seize it. It was also feared that the Maasai posed a threat to the railway; Nandi raids had made its forerunner, the Uganda (or Schlater's) Road, unsafe since the mid-1890s, and Nandi had attacked telegraph and railway survey parties. But in fact the Maasai did not follow suit. There are no reports of their ever having attacked travellers or gangers on either road or railway. In his railway-building days in the Rift, when he lived alongside the Maasai and bought milk and firewood from them, McGregor Ross did not describe any serious violence in his letters home. The most they did was snatch a white tablecloth from his tent one night, probably out of sheer curiosity.

Settlers had other concerns, articulated by Lord Delamere. He was the first individual to apply for a major land grant in the Rift, which was refused on the grounds of Maasai rights. He kept trying, and by

November 1903 he was given a 99-year 100,000-acre lease at Njoro, west of Nakuru, the first of several concessions. It cost him just over £200 a year.[15] Delamere said he foresaw a danger of clashes between warriors and settlers in this area. He suggested to government that the Maasai should be moved out of the Rift into a reserve and guaranteed no further disturbance there. He thought it a good idea to make this a game reserve, too, since the 'tribe' did not eat game and would not endanger it. That is exactly what happened on Laikipia, except that the guarantee proved worthless. It is very doubtful that Delamere actually feared Maasai aggression; his motive was the European monopoly of the highlands. As founder of the Planters' and Farmers' Association (which later reverted to its original title of the Colonists' Association), he was instrumental in pushing both for white reservation of land and the repulsion of Zionists. By early 1904 he was reportedly planning a private settlement scheme that the Foreign Office knew nothing about.[16]

The options: to mix or isolate

Eliot did not advocate native reserves, though he saw it might prove necessary to create one for the Maasai. He initially believed in intermingling Maasai and Europeans, on the grounds that assimilation would be better than isolation in reserves. Isolation would prevent the improvement of the 'race', and encourage the warriors to continue raiding.[17] 'But I quite recognise that the stupidity of the Masai or the brutality of Europeans may render it [intermingling] impossible and therefore we must have a reserve ready if needed,' he wrote in 1903.[18] Ainsworth told Eliot that the intermingling of black and white was feasible only if the Africans in question were cultivators, rooted in one piece of land. Nomadic pastoralists, he said, had insatiable grazing needs that were incompatible with those of European farmers, and they were monopolising the best pasture.[19]

At this time the Foreign Office favoured the idea of dividing the Rift Valley between Maasai and settlers. Officials Jackson (then Eliot's deputy) and Sub-Commissioner Stephen Bagge had suggested this when home on leave, while their colleague Charles Hobley drew an elaborate map showing how the carve-up would work.[20] Jackson told Secretary of State Lord Lansdowne: 'The Masai will never give us serious trouble so long as we treat them fairly and do not deprive them of their best and favourite grazing grounds, i.e., those in the vicinity of Lake Naivasha.'[21] Lansdowne had already told Eliot three months earlier: 'It is most important to avoid, so far as possible, giving them cause of complaint against the Administration.'[22] Jackson was concerned that new land applications would, if granted, devour 75 per cent of Maasai pastures in the Rift and on the Kinangop plateau, and urged the Foreign Office not to make any more grants (besides that given to the East Africa Syndicate, EAS) south of a line between the north ends of Lakes Nakuru and Elmenteita. He told Lansdowne that Eliot had recently assured Maasai elders that no further grants would be considered between Naivasha and Nakuru; Eliot later denied this account of the meeting.[23]

Hobley – who began 1904 as Assistant Deputy Commissioner and became Acting Commissioner when Eliot left – initially advocated leaving the Maasai temporarily in the Rift, concentrated in certain areas, and compensating them for the pastures they would relinquish to settlers. He said their flocks and herds had over time greatly improved the grazing and made it sweet; he thought it only fair that this fact be recognised. But soon afterwards, he and Ainsworth came up with plans for two permanent reserves. Crucially, the words permanent reserves were underlined in black ink in the original despatch. The accompanying map was marked: 'The red areas are proposed permanent reserves', with the last two words capitalised – but someone at the Foreign Office later ringed this sentence in pencil and wrote 'omit'.[24] There was another subtle change: Hobley now suggested that the proposed £3000 compensation for grazing should go towards the cost of removing and resettling the Maasai. In effect, they were to pay for their own removal. The costs were eventually met from savings and excess revenue, and there is no evidence that any compensation was paid.

Why did Hobley change his mind? He gave little indication, beyond saying that he had come round to the idea that 'the final solution' lay in settling all the Maasai on Laikipia, a view shared by Ainsworth and military chiefs. Hobley also claimed that the Naivasha Maasai had expressed agreement with the plan, without being promised a bribe, which removed one major obstacle – resistance.[25] One can speculate further. Hobley had only called the first plan (of mixing Maasai and settlers in the Rift) 'tentative', and would have been under pressure from all sides to find a permanent solution. Rinderpest had broken out in Maasai herds south of Naivasha in March, and about a month later near Nakuru, so the veterinary advice may well have been to get them out of the Rift and away from imported stock, though this was not spelled out in despatches. Also, there were growing fears of settler belligerence towards Africans, particularly from growing numbers of 'white roughs' from South Africa who were believed to be racist. Eliot wanted a small white police force established to keep them in check. There were lingering fears that the Maasai might attack whites if they remained alongside them; a military intelligence report in September 1903 began by saying the Maasai were unlikely to turn violent, but outlined plans for armed retaliation should this prove necessary.[26] There were concerns late that year that newly circumcised warriors were planning to launch a major raid in south-western BEA in order to blood their spears, but this was averted. Mungeam suggests that Hobley feared Maasai 'arrogance' could result in violence: 'His very reason for defending the move was that, had the Maasai been allowed to remain in their traditional grazing areas of the Rift, their arrogance would eventually have led to outrages and subsequent punitive expeditions.'[27] Hobley had seen how lovely Laikipia was when he inspected it (at Eliot's request) in June, recommending it as a reserve that offered 'magnificent grazing country' and water sources. He believed the government was 'deeply committed' to several potential settlers in the Rift, and could not back out now.[28] The Foreign Office was anxious to avoid expensive and embarrassing litigation, which would-be settler Robert Chamberlain was threatening to bring if he did not get his promised land (see below). The long absence of Jackson and Bagge at a crucial time meant that they could not influence events. Bagge had gone on leave in November 1903 and did not return until June 1904, when he was moved to Kisumu. Jackson went on leave shortly after Bagge; ill health prevented his return before February 1905. Lastly, the Foreign Office was fast losing patience with Eliot. When his face no longer fitted and resignation was on the cards, maybe it also rejected his antipathy to reserves and warmed to Hobley and Ainsworth's alternative plan.

Eliot oversteps the mark

Intermingling was not to be. By February 1904, the Foreign Office discovered that Eliot had exceeded his powers in promising leasehold grants in the Rift of 32 000 acres each, with a right to buy 10 000 of them at eight pence an acre after five years, to two British South Africans, Robert Chamberlain and A. S. Flemmer.[29] The two had reportedly lost no time in offering land in BEA on the Johannesburg market. After the EAS and Lord Delamere, they were respectively the third and fourth applicants for major land grants in the Rift. There was an official limit of 1000 acres on freehold grants, and Eliot should have sought Lansdowne's approval. Tipped off by Jackson and Bagge in London, the Foreign Office feared that Maasai rights had been ignored, overturned Eliot's decision and demanded an explanation.

Eliot was stung by the news that his underlings had been consulted behind his back. He was particularly hurt by the apparent disloyalty of Jackson. He believed he was contractually bound to Chamberlain and Flemmer, and refused to back down. He also felt that a precedent had been set by the home government, which in December 1903 had granted a 25-year lease of 500 square miles in the Gilgil-Naivasha area to the EAS for agricultural development. The Syndicate was a group of South African and City of London financiers, formed initially to prospect for minerals in BEA and Uganda. Eliot considered his own actions were consistent with this grant, and offered to resign on a point of principle. His resignation was eventually accepted and he had left by late June, after demanding a public inquiry. This never happened.[30]

In defending himself to the Foreign Office against Jackson and Bagge, Eliot launched a personal attack on Jackson in particular – calling into question his lack of native languages (how could he know what the Maasai said, when they would raid, or how much land they really required?), his favouritism ('Mr Jackson seems to think all Masai angels and all Somalis, Indians and Europeans devils') and his whole approach to governing: 'The root of the matter is ... that Mr Jackson is one of the strongest supporters of what I may call the gamekeeper theory of the Protectorate. He limits our task to protecting a few natives and preserving a little game. I have much sympathy with this easy and attractive theory of our duties, but it seems to me to have been demolished by the construction of the Uganda Railway and the expenditure of large sums for which some return is hoped.'[31]

In Eliot's third despatch to the Foreign Office in April 1904, defending his actions in the Chamberlain-Flemmer case, he went further. Eliot spoke frankly about the Maasai and his plans for their demise in this confidential letter which Sir Clement Hill, head of the African Protectorates Department, published in full. He began by questioning why the Maasai should get special treatment – 'The Masai are not essentially different from the Nandi, Lumbwa and other tribes. To regard them as especially friendly or loyal is unjust to other natives ... Neither can I see that they have any greater claim to land than other tribes and there seems to me something exaggerated in all the talk about "their own country" and their immemorial rights over which Mr Bagge waxes eloquent.' Then he launched a full-scale attack:

No doubt on platforms and in reports we declare we have no intention of depriving natives of their lands, but this has never prevented us from taking whatever land we want ... Apart from questions of expediency, justice does not in the least require us to reserve large tracts for the Masai; on the contrary, it would be an act of unjust partiality to treat them differently from other natives ... Your Lordship has opened this Protectorate to white immigration and colonization, and I think it is well that, in confidential correspondence at least, we should face the undoubted issue – viz. that white mates black in a very few moves ... there can be no doubt that the Masai and many other tribes must go under. It is a prospect which I view with equanimity and a clear conscience. I wish to protect individual Masais [*sic*] ... but I have no desire to protect Masaidom ... The sooner it disappears and is unknown, except in books of anthropology, the better.[32]

Now he had fully shown his hand, making no attempt to pretend that his administration would place African rights above European commercial interests. Hill used this confession to denounce Eliot to Lansdowne.

An exhaustive appraisal of the Eliot affair is not appropriate here. But from the bulky correspondence – published as Cd. 2099 – it is clear Eliot was at least no hypocrite, was consistent in his dealings, and had grounds to believe there was something suspect about London's rubber-stamping of the Syndicate lease. 'Terms of offer seem to me too easy,' he cabled.[33] Also, its property manager in BEA, J. K. Hill, happened to be Sir Clement's nephew John; though Sir Clement was not listed as a shareholder, he may well have been a beneficiary. The grant was certainly less defensible than smaller ones to individual settlers, both in terms of official obligations to pump in capital and develop the land, and flagrant abuse of Maasai grazing and watering rights. The Syndicate was not even asked to pay rent for the first seven years, and then only £500 per annum for the next 18. The grant appears to have been a reward for having spent £34,000 on mineral prospecting, though this proved fruitless except for the discovery of soda deposits on Lake Magadi. Eliot was right to rubbish the claim that the Maasai were happy about the grant, and had expressed their confidence in the Syndicate's local representative, Major Eric Smith. They clearly had no idea what he was up to, nor understood the long-term implications. What is also apparent from this file is the fact that the Foreign Office consistently emphasised the need to respect 'native rights' when considering land grants to Europeans. It finally capitulated to the local administration, after being reassured that the Maasai were anxious to move. In contrast to Eliot, Jackson frequently spoke up for Maasai land rights. Hobley, too, insisted that 'Maasai rights are a very real thing'.

Chamberlain exploded on hearing that Eliot's offer of land had been withdrawn. He railed against the 'squalid standard of British justice', and cursed his misplaced confidence in it. There were early lessons here for the Maasai. In a spate of letters he vowed legal action and revenge, claiming that he and Flemmer had done Britain a service by sending 150 settlers to BEA from South Africa and advertising its wonders there. This was all the thanks they had got. But when he learned of Eliot's resignation, he stopped berating the man and expressed 'deep regret [and] astonishment'.[34]

The first move

Eliot's successor, Sir Donald Stewart, moved swiftly to prevent further mishaps of this kind. On arrival in BEA in August 1904, he went straight upcountry to find out what was going on in the Rift: 'Sir Donald ... came to the conclusion that the removal of the Masai from the Rift Valley into two reserves was the only real settlement to the question. Masai and Europeans could never live together without endless trouble and friction.'[35] Hobley and Ainsworth had already submitted their plans for reserves to London, drafted a treaty and allegedly gained the verbal agreement of Maasai leaders at Naivasha and Olonana at Ngong. Settlers Chamberlain and Flemmer got their land, the Colonial Office telling the former: 'The decision to remove the Masai from the Rift Valley disposed of one of the most serious obstacles to allowing you and Mr Flemmer to receive a suitable grant in the localities which you originally selected.'[36] The relief was mutual; litigation had been avoided. Stewart alluded to this threat, saying of the Rift, 'the whole of it is really Masai country' and the leases could only be granted if the 'tribe' was removed. Otherwise the government would be obliged to offer these settlers land elsewhere, and if they refused it, 'fight the matter out in the Courts'. If it came to this, he doubted Britain would win.[37]

This personal victory did not stem the flow of Chamberlain's complaints to government. He continued to rant about the preferential treatment shown to financial syndicates and aristocrats, calling the EAS a bunch of gamblers and mineral prospectors hiding behind a sham front of 'harmless farmers and innocent graziers'.[38] He had a point. Once Eliot had gone, he and Flemmer were offered only 20,000 of the 32,000 acres originally promised. Flemmer accepted, but Chamberlain was already occupying the land Eliot had offered him and refused to let surveyors cut out the difference, threatening to shoot any who set foot on his land. He got his way eventually, after appealing to Winston Churchill.

Lansdowne gave his blessing to the 1904 Maasai Agreement, signed in August, though he expressed surprise at the speed with which Stewart had rushed it through.[39] It is unlikely that this was due to Stewart's capitulation to settlers; they did not have a cosy relationship. According to Hollis, 'he stood no nonsense' from them. Lord Delamere had confided: 'he never knew when he went to Government House whether the ADC would be instructed to give him a whisky and soda or to kick him down the steps!'[40] The Maasai signatories were Olonana, Masikonde and 18 representatives of eight sections, 12 of them *il-aiguenak*, including Ole Gilisho. This is important, because it implicitly recognised the authority and representativeness of the age-set spokesmen. The Loitai were said to be represented, but none actually put their mark to this document. The signatories agreed to vacate the whole of the Rift Valley and move into two reserves – linked by a connecting road – which were promised in perpetuity. The Purko, Keekonyokie, Loitai, Damat and Laitutok (Laitayiok?) sections were to move to Laikipia, while the Kaputiei, Matapato, Loodokilani and Sikirari were to move south. A sacred site on the Kinangop in the Aberdares, where circumcision and other ceremonies were traditionally held, was to be retained for Maasai use.

Most significantly, in the light of what happened later, British officials stressed the permanence of this arrangement. Lansdowne 'emphasised the fact that the definite acceptance of the policy of native reserves implied an absolute guarantee that the natives would, so long as they desired it, remain in undisputed and exclusive possession of the acres set aside for their use'. (Churchill repeated the point when speaking of the inviolability of native reserves in the House of Commons in July 1907.) When he submitted the treaty, Stewart had prophesied that settlers might soon cast 'envious eyes' on the Laikipia pastures once Maasai stock had grazed the grass down and 'got it sweet' – just as they had in the Rift. For this reason, he urged the 'absolute necessity of making these Laikipia lands an absolute native reserve for the Masai'. The likelihood of settler covetousness was raised at a Foreign Office meeting to discuss the settlement proposals. Here it was also noted that 'the settlement should be looked upon as of a permanent nature, i.e. that if in the future any change became necessary the Masai should be entitled to compensation'. Lansdowne minuted on this in red: 'Surely if they are to be moved they should be settled permanently in their new home.'[41]

No one moved quickly. In his unpublished autobiography, Hollis said that when he returned to East Africa in January 1905, the Maasai had already been moved. But the first mention in the PRO registers is of Maasai moving by February, and they were still doing so in June. Stewart wrote: 'A station has been selected [on Laikipia], and has received the name of Rumuruti, materials are being rapidly transported for the erection of the necessary buildings, and the Masai themselves are removing their families and stock to their new homes quite readily and contentedly.'[42] There was no mention of resistance. However, Sandford admitted, quoting Hollis, that 'some pressure had to be put on the Masai of the Rift Valley to induce them to leave their grazing grounds'.[43] Leys wrote that the Rift 'was most unwillingly evacuated'.[44] An anonymous letter writer to the Anti-Slavery Society claimed: 'Masai very loth to leave. Villages burnt by Government.'[45] DC Collyer's interpreter Marianyie Ole Kirtela, interviewed by Hanley, attended a public meeting with the British at which the move and treaty were agreed. He let Hanley in on a 'secret' – Olonana, who had been salaried by the British since 1901, had already agreed to both move and treaty in advance of the meeting. 'Elders were there in hundreds, most of them from the Purko section,' said Ole Kirtela. 'But the Purko section refused to leave Naivasha and Kinopop [Kinangop] and go to Ngatet. If they were asked to move up into Endoror they might reconsider the matter. They protested and said that Ngatet was too far.' The Keekonyokie also lodged objections, refusing to move either to Ngatet or Entorror, but they were overruled.[46]

Why did the Maasai not rise up at this point? They were in no position to resist because they were in recovery, still reeling from the successive devastations of disease, drought, famine and internal disorder at the end of the nineteenth century. Their stock losses to rinderpest alone are estimated to have been as high as 90 per cent.[47] Many had been forced to seek refuge with neighbours, including the British at Fort Smith. They also knew, having taken part in punitive expeditions, that their spears were no match for British guns. They could not resist without Olonana's sanction, and it appears he was not averse to the move. The only dissidence that showed itself in 1904–05 was failure to do what they were told; many Maasai simply did not go to their allotted reserve. According to Sandford, about a quarter of the Purko stayed in the south instead of going to Laikipia. The Loitai continued to occupy the Loita hills and plains; the only reason why small numbers had gone to live in the Rift at all was because some had accompanied Senteu there. The Damat and the Laitutok also stayed south, while the Keekonyokie moved a short distance from Naivasha to Melili and the southern slopes of the Mau. However, some Keekonyokie stayed on around Naivasha; there is a later reference to their move from Naivasha to the Loita plains in 1910.[48]

The first move requires more research. But Maasai oral testimony typically does not distinguish between the loss of the Rift and the loss of

Laikipia. Most of my informants referred to the whole of their former northern territory as Entorror, and to that portion of the south (now western Narok) to which they first moved as Ngatet. From now on, all references to Entorror will implicitly include land occupied pre-1904, but most recently Laikipia, unless stated otherwise. The word Entorror comes from the verb *a-rror* meaning to trip or fall down. Informants said this referred to the environment: 'It is such a good place, that one becomes spoilt and does not last long.' Waller gives a different interpretation: that the central Rift had been a focus for hostilities during the Iloikop Wars and therefore 'falling down' referred to people being killed.[49] I find this unlikely, given the fact that Entorror encompasses a much larger area, including Laikipia, and is today primarily remembered for its 'sweet' environment (see Chapter 5).

The Laikipia experience

For the most part, the Maasai prospered on Laikipia, despite periodic outbreaks of bovine pleuro-pneumonia (BPP), 'gastro-enteritis' (likely to have been rinderpest), and other stock diseases which will be covered more fully later. The Purko in particular had already come out of late nineteenth-century inter-sectional warfare and *emutai* in the best shape, compared to other sections, and they consolidated their gains from 1904 to 1912.[50] Between 1904 and 1911, Maasai stock as a whole in both reserves 'probably trebled' according to Leys. In fact, official figures for the Northern Reserve indicate that cattle more than trebled in just five years – from 64,000 in 1906 to 200,000 by 1911, and the latter figure does not include the 10,000 cattle that went south in early 1910 after the *eunoto* ceremonies on Kinangop.[51] Waller writes of the 'highly productive pastures' on Laikipia, where the Purko rapidly expanded their stock and probably became wealthier than their kin in the Southern Reserve.[52] This wealth was not simply held in cattle; small stock is the bedrock of the pastoral economy and often provides a safety net when food is scarce. Sections such as the Purko and the Loitai, which owned large numbers of sheep and goats, were therefore at an advantage immediately after *emutai*. The number of sheep on Laikipia was estimated to be 1.75 million in 1906.[53]

The reserve was too small for their needs, and was extended twice. The Maasai frequently broke out to find more grazing during droughts, and were sometimes fined. In the east, they were permitted to cross the Uaso Nyiro river every dry season. In the south, some settled by Lake Ol Bolossat. Stewart's successor James Hayes Sadler (Stewart had died in

October 1905, and Sadler arrived by December) asked the EAS to give up its lands there to the Maasai, but no agreement was ever reached. When the rains partially failed in 1907, the Maasai were again allowed to break their boundaries in the east, south and west. The grazing was 'very scarce' in early 1910, and the Maasai broke out of the reserve all along the Uaso Nyiro, incurring convictions and fines. This confirmed the opinions of both Bagge and Collyer that the reserve was not large enough for their needs, but Bagge's 1908 proposal to extend it again was not acted upon.

Olonana never visited the 'northern' Maasai on Laikipia after 1904, and his authority waned. It had been tenuous at the best of times. The northern faction had effectively broken away from his control, if indeed this had ever existed, following a trend that began at Naivasha, where Jackson and MacAllister (the Collector there) had pursued a policy of building up Ole Gilisho's power to rival that of Olonana. Purko on Laikipia in 1902 told administrative officer H. R. (Harry) Tate that they acknowledged Olonana as their 'chief' but did not consult him in any way. At the next major circumcision ceremonies, the section was divided by arguments over the choice of age-set spokesman for I-Lemek, the new left-hand circumcision group of Il-Tuati II (Il-Dwati), the juniors to Ole Gilisho's warriors. Olonana favoured Ole Goinyo, while the 'northern' Maasai chose Ole Kotikosh. When the reserves were created, Ole Goinyo and his supporters went south while Ole Kotikosh and his people went to Laikipia. The row simmered for years, coming to a head again in early 1910 when the *eunoto* ceremony was scheduled to take place on the Kinangop. 'It is a great pity that Lenana will not come to Laikipia, for it is hardly to be expected that he can maintain his influence here with people he never sees,' wrote Collyer that year. Another reason for Olonana's failure to visit Laikipia may have been his poor health, which steadily worsened in the year or so before his premature death in March 1911.[54]

On Laikipia, younger leaders of the Il-Talala and Il-Tuati age-sets, notably Nkapilil Ole Masikonde and Ole Gilisho respectively, were able to consolidate their power base, free from Olonana's interference. They were already in a strong position. The balance between age-sets had been upset by the events of the late nineteenth century. A stratum of older men had been taken out by the disasters, which handed an unusual degree of power to junior elders and senior warriors of Il-Talala and Il-Tuati. Their stock wealth had grown as a result of their participation in British punitive expeditions, as well as raiding on their own account. 'As a result,' writes Waller, 'they reached power and prestige earlier than usual and kept it for longer.' In particular, the right-hand circumcision group of Il-Tuati was a supremely self-confident bunch, whose confidence and authority stemmed from the fact that they had repelled the Loitai, and restored the community's depleted herds through raiding, sponsored and otherwise. They also enjoyed reflected glory, since their sponsoring elders (known as fire-stick patrons) were the famed Il-Aimer, who had shone as warriors before the disasters struck. As one of Waller's informants put it: 'It was said that Il Dwati [sic] brought the Maasai back to life'.[55]

Collyer reported in 1910, in a foretaste of what was to come:

> The two biggest Chiefs in Laikipia are Masikondi and Legalishu (Ol Le Gilisho). Of Masikondi I have nothing but good to say; he is most helpful in every way ... He has seen enough of the white man to realize that the Masai must move with the times and he is prepared to progress and use his influence for forward movement ... Unfortunately his influence for progress is largely discounted by the influence of Legalishu who takes his cue from Lenana, and is against change of all kinds.

Ole Gilisho was considered 'intensely conservative', but his influence was believed to be greater than that of Masikonde, and he was 'far too big a man to be ignored'.[56]

Girouard engineers the second move

By 1908, the administration was already hatching plans to move the 'northern' Maasai to an extended Southern Reserve in defiance of the 1904 Agreement. At this stage, the Colonial Office was not informed; the first it knew of these plans was when it was shown the budget estimates for 1910–11. Local officials were perfectly frank, at least in the district records in which these plans were described, about what lay behind a possible second move: 'As soon as Laikipia is free from Masai the district will be open for white settlement.'[57] This reason was not publicly admitted; the official line was that the move was for the benefit of the Maasai themselves, and that Olonana demanded this.[58] Tours of inspection of the proposed western extension of the Southern Reserve were made by Herbert McClure (then ADC Southern Reserve) in November 1908, and by Hollis and Bagge in April the following year. They reported a lack of water, saying this could be remedied by irrigation. Watts, Commissioner of Public Works, drew up plans for this on Loita.

But was there another motive, linked to settler pressure on the government to take action on East Coast fever (ECF)? Initially called African Coast fever, it was first diagnosed in the Protectorate in 1904, in a herd of cattle brought from the Kilimanjaro area of German East Africa to Nairobi.[59] It hampered the earliest white settlers' attempts to set up dairy and beef ranches; Delamere lost nearly all his young stock to it at Njoro. Settlers were panicked by news of how ECF was devastating Rhodesia and the Transvaal, following outbreaks that began in 1901 and 1902 respectively.[60] Churchill, who visited BEA in 1907 as a junior Colonial Officer minister, was briefed about ECF and suggested remedies – wire fencing and quarantine – in his account of that journey.[61] It was thanks to him that the Treasury released extra funds so that farmers could buy fencing material. By 1909–10, ECF was seriously worrying settlers who in turn harassed Governor Percy Girouard, who had replaced Hayes Sadler in the autumn of 1909. Sadler had had a poor relationship with the settler community, and Girouard wanted to make amends, never mind friends. He was sufficiently concerned about ECF to wire the Colonial Office on behalf of individual settlers facing outbreaks on their farms, who were demanding more land in 'clean' areas (see Chapter 5). In April 1910 he wrote to H. J. Read, head of the East African Department in London, on behalf of a Mr Heatley 'who has had very bad luck owing to ECF' on farms at Kiambu and wanted 'extra land in a district unaffected by this pest'. He asked Read to intercede with the Secretary of State. Three days later, he wired Crewe to report another outbreak of ECF on Heatley's farm.[62]

These actions seem odd: had a governor not got other things to do? According to this correspondence, the highlands were free of ECF at this point. In February 1910, Girouard wrote to the Secretary of State begging for more money to tackle ECF in order to prevent it from spreading to the highlands. Confirmation of the situation on Laikipia came in February the following year, when Acting Chief Veterinary Officer Francis Brandt visited Laikipia to investigate an outbreak of pleuro-pneumonia and to 'find out if ECF is known on Laikipia'. He drew a blank, according to Collyer: 'He could find no trace of ECF and information supplied him by the Masai pointed to the fact that although the Masai feared the disease and *knew it existed in the South*, no case had *ever* been known on Laikipia'.[63] (My emphasis.) Plans to move the Maasai were immediately accelerated, which was surely no coincidence.

Of course there were also other reasons for the move, which would have happened sooner or later anyway. As land-hungry settlers made growing demands for farms in the highlands, the relocated Maasai came to be seen as a moveable embarrassment. Seven people had applied for land on Laikipia before 1904. When the first Agreement was signed, they were offered land elsewhere as compensation for giving up these claims. All but two did so; Delamere and his brother-in-law Galbraith Cole eventually succeeded (in April 1910) in persuading the Land Office to recognise their original claims.[64] Furthermore, the Maasai's physical division into two reserves did not make administration easy. It was impossible to control stock disease while herders trekked to and fro between them, defying quarantine, and reserving the right to congregate both stock and humans in their thousands on ceremonial occasions. Seen from a new administrator's perspective, the whole idea of two reserves had been quite mad in the first place.

Percy Girouard was a French Canadian married to an English South African, Gwen Solomon, daughter of the Attorney-General of the Transvaal. He arrived in BEA from Northern Nigeria, where he had earned high repute as Lord Lugard's successor. An engineer and graduate of the officers' course at Britain's Royal Military College, Girouard was essentially a builder and director of railways – in the Sudan, Egypt and South Africa.[65] He was Director of Railways for the British army in South Africa during the South African War, before becoming governor of a railway company in the Transvaal, and Commissioner of South African Railways. In the second of these South African posts, he became used to getting whatever he asked of his employers, no expense spared. For example, after persuading Lord Milner to lobby Colonial Secretary Joseph Chamberlain, he was given £1 million to buy new locomotives and other goods. His dismissal as Commissioner of Railways did not dent his confidence. By the time he reached northern Nigeria in 1906, he had acquired a real taste for power, writing to tell his father: 'excepting the hold of the Secretary of State, I am a little independent king'.[66] It was a style of command he would use in East Africa.

Girouard had the air of a New World frontiersman. He soon found the desk-bound dictates of the Colonial Office irrelevant in a country where he believed 'the man on the ground' should set the pace and the policy. Where his predecessors had taken care to separate imperial duty from settler interests, at least in public, Girouard alarmed London by being unashamedly pro-settler. Like Eliot, he rapidly succumbed to Delamere's influence. He actively promoted settler interests and tried to cut some of the red tape ties with home, pushing for self-government. In an unpublished manuscript, he called for settlers to be given 'freedom in the management of their own affairs'. On the other hand, his biographer suggests that Girouard was not unsympathetic to African rights.

He advocated indirect rule because he believed it prevented African authority from being undermined. Before coming to BEA he had studied land tenure in other parts of the empire, and favoured 'the definition and recognition of native rights in land'.[67]

But he had inherited a problem with the Maasai. They now occupied a superb stretch of country on Laikipia that white settlers coveted, but which had been promised to them for life. How could they be made to 'disappear', as Eliot once suggested, without sparking rebellion at home and abroad? As Sorrenson explains: 'It was a measure of the utmost delicacy. Girouard had to abrogate the 1904 Masai treaty and pretend to the Colonial Office that the Masai wanted to move south. At the same time he had to disguise the fact that he was acting in the interests of the settlers, some of whom had been promised land on Laikipia.'[68] It is also fair to stress, as Cashmore does, that Girouard was not the chief architect of the second move. It had been on the cards for some 18 months before he arrived in BEA, and his main aim in this and other sluggish administrative matters was to shake things up and move forward.

Without waiting for a formal treaty or Colonial Office sanction, Girouard began to move the Maasai south in early 1910. He used the impending *eunoto* ceremony on the Kinangop as a cover. This was scheduled to take place on a sacred site between the two reserves, as allowed under the 1904 Agreement. In December 1909, Purko warriors from Laikipia were permitted to proceed to the Kinangop with 10,000 cattle, accompanied by Ole Gilisho and Masikonde. Maasai on Loita were told they could also attend, but they would have to leave their stock behind because of the risk of infecting settler stock *en route* with ECF. In January, Olonana kicked up a fuss – ostensibly over threats to his authority posed by the division into two reserves, but more likely linked to the ongoing row with 'northern' Maasai leaders over the choice of age-set spokesman. He demanded that the ceremony be switched to Ngong, where he lived, in the Southern Reserve. At first, Ole Gilisho and Masikonde refused. Girouard had a private meeting with Olonana on 2 February to thrash the problem out, telling the prophet not to let the others know what they discussed. Girouard later told the Colonial Office that Olonana had, at this meeting, asked the government to reunite his people in one reserve, and that it was vital to support his authority. He claimed: 'The whole matter has really been a demand from the chiefs themselves who are influenced in no way by anyone.' In fact, he had struck a private deal with Olonana, recognising him alone as Paramount Chief in return for his cooperation in getting the Maasai to move.[69]

It is unclear whether the *eunoto* went ahead – Sandford said it began prematurely on the Kinangop, without Olonana's sanction, and after the talks with Girouard all those involved moved south. Sorrenson says the cattle were driven straight to Loita, and the ceremony was postponed. Either way there is an obvious anomaly here: if ECF quarantine was the reason why the 'southern' posse could not come north, how come there were no government objections to 'northern' cattle moving south through European farms? Clearly ECF was just an excuse, or the government believed the Laikipia herds were free of it and therefore posed no threat to settler stock. These migrants were intended to be the vanguard of the entire Maasai population of Laikipia.

The Colonial Office received a tip-off about Girouard's plans when Gilbert Murray forwarded a letter of warning from Leys, omitting Leys's signature.[70] He accused the government of acting unjustly and breaking its pledge to the Maasai, and ended by telling Murray: 'You can depend on the accuracy of my account. Act with regard to it entirely on your own discretion. Use my name if you like, if [his underline] it makes any difference of success. I don't see how it can. And it would probably mean my dismissal.' Sorrenson claims no official notice was taken of this letter. In fact, the move was called off on 19 April by telegram, as he himself describes a few pages later.[71] Leys was surely instrumental in this, and in sowing the seeds of doubt about Girouard at the Colonial Office, which led to his ultimate downfall. The Governor was told to leave the Maasai where they were until such time as they chose to leave, and not before another treaty had been negotiated. But he made sure that potentially obstructive local officials with Maasai sympathies were removed by one means or another – notably Bagge (by now PC Naivasha) who resigned in March, ostensibly for health reasons.

Key meetings

The first of a series of key meetings to discuss the proposed move was held with the Maasai on 24 February 1910 at Kiserian camp, close to Olonana's homestead at Ngong. It was here that Jackson, in Girouard's absence, told the 'northern' Maasai what had already been agreed with Olonana. He, McClure (Acting DC Southern Reserve), Lord Delamere and Collyer met Olonana, Masikonde, Ole Gilisho, Saburi (described as Olonana's chief elder) and others. Jackson urged relocation in the south out of concern for the 'safety' of the Maasai. Most significantly, he defined safety as freedom from cattle disease: 'the Masai in the Southern Reserve were safe, and the Purko would be equally safe if located in the

Loita country.' He was aware that the Maasai had always kept their promises, he said, and the government wished to keep theirs. The move was being considered in 'the best interests of the Masai and not because [the government] wished to take back the land already given them'. The three so-called chiefs all reportedly said they were in favour of the move, Olonana adding that he had explained to Ole Gilisho and Masikonde what country the Purko would be allowed, and that he had arranged for the *eunoto* to take place in the Southern Reserve. Olonana was said to be 'very pleased' with these arrangements. Ole Gilisho sounded more doubtful, asking for unspecified details to be clarified. He was reassured that these would be taken care of.[72]

A second treaty was agreed at a meeting in May at Ngong, but never implemented. It was signed by Olonana, Ole Gilisho, Masikonde and 11 other Maasai. Girouard did not sign it. Ole Gilisho, the one dissenting voice, made it very clear that he now opposed the move, because the proposed territory was not large enough and lacked sufficient water.[73] Before attempting to enforce the treaty, Girouard therefore invited Maasai representatives to inspect the proposed western extension of the reserve. Ol Le Geli (*sic*, probably Yiaile) and Reien represented Ole Gilisho, while Olonana also sent two men to accompany Collyer on the two-month trip. They 'pretended not to be pleased with the country', said there was not enough water, and doubted the government's ability to make dams to conserve sufficient water.[74]

Collyer called another major meeting to discuss the move with the Maasai on 27 August 1910 at Rumuruti.[75] The mood had radically changed since February. Collyer told them quite frankly they would 'probably be squeezed out of Laikipia', that it was not big enough for them anyway, and now was the time to unite with their kin in one area. The Maasai were equally blunt. They said they would move if the government ordered it, but they did not want to because the land was bad and waterless and the Sotik would steal their stock. Leaders were invited to inspect the proposed extension. Their response to this was taken down verbatim by Collyer, who said in a covering letter that the District Clerk's attempt was not quite what he wanted.

> We don't want to send anybody to look at any part of the country, between us, we know all of it ... We know the country offered is bad and waterless and have no faith that Government can store sufficient water for us. We are sure our stock will die there, but we are prepared to obey the orders of Government and go. We want Mau, but if we can't have it we shall have to do without it. We don't want any more

conferences on the matter, as the above is the decision of us all, we will go.

This was agreement of a kind but, as Collyer said, 'with a very bad grace'. He expressed surprise at the new truculence: 'The line the Maasai now adopt is probably the result of much thought. No inkling of what they proposed to say reached me previously.' He doubted these were minority views, and whether his superiors would accept the result of the meeting without calling another conference with Olonona, presumably in the mistaken belief that the prophet could push the dissenters into line. He sent the report of the meeting by special runner to the Governor via the Provincial Commissioner on 29 August 1910, saying 'it is hoped your answer will be returned in like manner'. He wanted to know what he should do next. This should have been forwarded to Lewis Harcourt, the new Secretary of State for the Colonies, but no one in authority was in a hurry to convey news of a Maasai change of heart.

In his August 1910 'Report on the Masai Question', Collyer advised against moving the Maasai again unless they would be better off with the grazing in the south, or at least no worse off.[76] His intention in this paper was to 'discuss the question from the point of view of the Masai, as distinct from the Settlers' "view" '. He described the 'well behaved' Maasai as never having presented a threat or nuisance to European settlers, the railway or its staff. They had always been cooperative, and moving them again would be a poor reward. He urged the government to view with suspicion Olonana's desire to 'unite' the tribe in one reserve: 'Lenana's motive is purely a political one, as he wishes naturally to have all his people under his own eye.' He himself would not have to move, nor would his stock suffer. Collyer urged the government to give the Purko some say in the matter if it seriously planned to move them again. As for white settlers, they had not wanted Laikipia originally. In his view there was every likelihood that they would want the Loita country, too, in a few years' time.

Collyer suggested alternatives to moving them, which included expanding the Northern Reserve, curbing stock numbers and raising taxes 'so that at any rate the Masai pay for their own administration'. He warned against trying to turn them into agriculturalists; the most likely way of making them 'useful members of society' would be to improve their stock. The report was not sent home immediately for the obvious reason that it contradicted the Governor's line. Harcourt eventually saw these and other reports by Collyer in March 1913, when their author was long dead.[77] Enclosed in the same despatch was Collyer's report of a meeting with Maasai leaders at Gilgil on 18 November 1911, when Ole Gilisho said he had only agreed to give up Laikipia under threat of deportation by administrator C. R. W. Lane. The age-set spokesman also said he had now personally inspected Trans-Mara but did not think it good enough. Collyer was dismissive, saying Ole Gilisho had only spent four days there. He agreed to take six of Ole Gilisho's representatives to inspect the area, and told them to visit with an open mind, not with the aim of making the worst of everything.

The Leys campaign

Norman Leys began writing to MP Thomas Edmund Harvey about the Maasai on 17 October 1910. Known as Ted to his friends, who included my grandfather John Hughes, Harvey was Liberal member for his home town of West Leeds from 1910 to 1918. An Oxford-educated Quaker, Harvey fought for penal and educational reform at home and colonial reform in India and Africa. He went on to defend the rights of conscientious objectors and enemy POWs in both world wars. The unfashionable underdog was his speciality; he was, therefore, open to hearing what Leys had to say about the Maasai.[78]

To briefly set this exchange in context, in the same period agitators were calling attention to colonial excesses elsewhere in Africa, riding on the momentum of the anti-slavery movement. Both men were aware of, and Harvey was at this stage certainly involved in, these broadly Christian socialist anti-imperial movements, through the Anti-Slavery Society and friends such as Gilbert Murray and Ramsay MacDonald. In the Congo, E. D. Morel and Roger Casement were speaking out against atrocities being carried out in the name of King Leopold. In West Africa, British investigative journalist Henry Nevinson was exposing Portuguese use of mainland slaves as 'indentured labourers' on cocoa plantations in the islands of São Tomé and Principe years after slavery was abolished, and thoroughly upsetting Quaker cocoa manufacturers who were shipping the fruits of this labour to an unsuspecting Britain.[79]

Contemporary African-led movements included the Aborigines' Rights Protection Society, formed in the Gold Coast in 1897, which sent delegations to London in 1898, 1906 and 1911 to protest directly to the Colonial Office about various new laws. The pages of the *Anti-Slavery Reporter*, from its first issue in October 1909, were full of news of São Tomé cocoa, native policy in Nigeria, rubber slavery on the Amazon in South America, lynchings and 'the race problem' in the United States, and the colour bar in South Africa. By 1911 this publication, produced by the Anti-Slavery and Aborigines' Protection Society in London, was flagging up the Cole case (see next chapter) and the Maasai moves.[80]

Leys was clearly inspired by Morel in particular, who also focused on economic causes and saw how the theft of African land and labour made Leopold's whole system of exploitation possible. Cell mentions: 'At the height of the Masai affair he [Leys] had compared the scandal of the East Africa Protectorate to that of the Congo several years before. Only gradually did the notion begin to crystallize in the mind of Norman Leys that he himself might become the E. D. Morel of East Africa.'[81] Short of becoming a journalist like Morel, he followed Morel's deft use of every available means to expose injustice: the press, books, articles, pamphlets, speeches and copious letter writing. Leys was aided by anonymous sources inside the system, just as Morel was inside its Belgian colonial equivalent. Though the Harvey Letters do not name Morel, Leys clearly feared that a Congo-type scandal might develop in East Africa when he declared of the Maasai issue: 'If the CO does not find out the guilt and punish the guilty there is a probability in a year or two of a large administrative scandal comparable to the Congo affair.' In 1913, he criticised the Anti-Slavery Society for failing to do for British colonies what it seemed able to do for the Congo and Angola. Later, Leys worked closely with Rev. John Harris of the Society – who had collaborated with Morel in the Congo Reform Association – while sitting on the League of Nations Union Mandates Committee. All these networks were interlinked.[82]

Furthermore, Leys's profession gave him unique access to places (such as prisons, steamships, workplaces and slums) and events (he accompanied a 1908 British punitive expedition against the Kisii) that yielded inside information he could use against the government. Besides the Maasai cause, he took up several others simultaneously and wrote in July 1912 that he was 'strongly tempted to take up a still more scandalous business than the Masai – the condition of contract labour'.[83] Curiously, Girouard also wrote to Morel to enlist his help in case he faced agitation by 'sentimentalists' in Britain. He confided in 1911: 'I have a small native question which is being criticised particularly by Ramsay MacDonald. I wish he would come and govern for a while … I have croakers who certainly don't know a Masai from a Fulani howling what a cruel beast I am.'[84]

This is the background to the barrage of letters that Leys now sent to Harvey and influential others. Leys, home on leave, had met Harvey in London the day before he sent the first letter, and already regretted some of what he had said. He wrote: 'If I had to live yesterday afternoon over again I would put things to you in a different way. To me this move of the Masai has been a sore business for over a year. And I am apt to speak of it out of proportion.' He felt powerless to do anything about it within the Protectorate, and called for a new, liberal colonial policy 'that must come from home'. He urged Harvey to help create this and avert disaster, for he believed 'we are laying the foundations of an evil system that generations to come will struggle with. Be ready in ten years for native "rebellions". I could write a fairly accurate account of one now!'[85]

His chief informant inside the administration was clearly Collyer, whose heart was not in the move. Cashmore describes how he was 'devoted to the Masai', having first administered them at Naivasha.[86] Another key informant was probably game warden George Goldfinch, whose role will be described later. Though he was appointed Special Commissioner in charge of the move, Collyer went on home leave from April to November 1911 and missed the first half. It is not clear whether he was deliberately removed. However, he refused to join Leys in public protest because he still believed in the good intentions of the British government.[87] Leys emphatically did not. Moreover, the concerns he shared with Harvey throughout their correspondence were broader than Maasai rights. He questioned the fundamental values and actions of government in a country dominated by settler toughs. For example, since 1908 he had discussed with McGregor Ross and others the possibility of issuing a pro-native manifesto or 'counterblast', as Ross called it in letters home. This was intended as a response to the worst excesses of settler misbehaviour and judicial indulgence of it: 'We want to show some of the South African stiffs that there is a community in the country flatly opposed to almost everything that they advocate.'[88] It was apparently never written.

In early 1910, Leys had unsuccessfully tried to get Collyer, Ross and a few other unnamed officials to agree to resign in protest at the Maasai move. Ross told him to first write to the Governor. Leys sent Ross a draft of his letter to Girouard: 'It was much too strong and undiplomatic. I largely recast it and sent it back to him today ... .' Leys sent the letter to Girouard at the end of May, and as a result was summoned to meet him. Girouard appears to have charmed Leys into uncharacteristic submission, from what he told Murray:

> The main impression left on my mind by the Governor is that he is better fit to be responsible for native rights and interests than any Governor I have served under. He told me he is dead against any other move of a native tribe. I blame him for having given settlers a

wrong impression of his policy – an impression I shared myself. He assured me that six months from now settlers would know the truth. On the Masai move he is determined and I feel that however regrettable that is, my duty is to say and do no more than I have, now that I know he is no friend to land grabbers. He has been misled into a mistake ... Sir Percy was so frank as to tell me his detailed policy. I could put my name with pleasure to nine-tenths of it.[89]

Ross did not hold out much hope of success, telling his mother: 'The facts are ... white settlers badly want Laikipia. They do not want the Southern districts near the German border. These have been open for settlement for some years and practically nobody will look at them. Therefore, clear the Maasai off Laikipia and send them down to the region which nobody wants! Delamere is a prominent advocate of the move, of course. Naboth's Vineyard.'[90]

Ross also spoke directly to Girouard about the Maasai, while visiting Government House on unrelated business. Girouard raised the subject of the move, saying he understood that Ross opposed it. Ross was unimpressed with Girouard's defence. 'He made out no case which a man of Leys' knowledge and ability could not counter effectively if he started writing in the home papers. I told him that I was naturally on the side of the evicted, that an aunt had been laid out by British soldiery from Inverness in resistance to the depopulation of a glen that was wanted by people of superior standing, and that my sympathies automatically went in that direction.' He clearly saw parallels with the highland clearances. Ross advised Girouard to check that there was sufficient water in the Southern Reserve before concentrating five million head of stock there.[91]

Leys did not write to Harvey again until May 1911, a month after becoming Acting Medical Officer at Mombasa. He had gone on home leave in February, and apparently briefed Harvey and Ramsay MacDonald face to face. 'Things have gone badly again,' he wrote:

The Governor again proposes to move the Masai. This time he has engineered a request from the people themselves. I don't know the whole story ... How can such intrigues be proved? To accuse Girouard of them would simply be to procure my own dismissal. He is far too crafty a general to allow an outsider like me to know what goes on. I believe he will fall. I don't think Providence often allows the wicked to prosper very long. But I fancy he is much more likely to hit his own wicket than to allow himself to be bowled out.

Leys urged Harvey to ask more questions about the move in the Commons.[92]

A 1911 Maasai Agreement had been signed the previous month.[93] It negated what had been agreed in 1904. Under this new deal, the British sought to 'unite' the Maasai in one extended Southern Reserve. Some 4500 square miles on Laikipia were to be exchanged for 6500 in the south, and this time round there was no talk of a permanent contract. It was signed by 15 Maasai (or rather, they allegedly set thumbprints next to their names), led by Seggi, son of the late Olonana. Of the 15, only Ole Gilisho (now one of Seggi's regents, since he was a minor) and Masikonde had also signed the 1904 Agreement. Girouard told London that the Agreement had been sanctioned by Olonana. On his deathbed, on 7 March 1911, Olonona was said to have exhorted the northern Maasai to obey the government and move south. Such dying wishes were a godsend for the Governor, who pressured Olonana's successors to sign the 1911 Agreement days later. The timeliness was not lost on Harcourt, who dryly noted: 'Sir P. Girouard must have telepathetically inspired Ole Lenana's dying speech!'[94] Subsequent events showed that Girouard's confidence in Olonana was misplaced, and Olonana's authority highly questionable. Olonana reportedly died of dysentery, though rumour persists among the Maasai that he was poisoned or bewitched by his brother Senteu. Edward Crewe-Read (then ADC in charge, Southern Reserve) spoke to him just before his death, viewed his body a day after he passed away, and recorded: 'Thus ended the life of one of the most powerful and intelligent natives this country has known.' The source of the 'dying wishes' story was attributed to his half-brother Marmaroi (or Marmoroi), whom elders recommended should be Seggi's chief advisor until he came of age.[95] That advice was rejected, and Ngaroya (Nkaroyia, Olonana's cousin) and Ole Gilisho became his regents.

An anonymous informer again threatened to scupper the contract. Colonial Office approval of the 1911 Agreement was stalled by Ramsay MacDonald, then MP for Leicester, who passed on information from BEA (supplied by Leys) alleging that Girouard had forced the Maasai to sign. MacDonald reckoned that Olonana's dying wish had been 'manufactured by the living', and that settlers were trying to provoke a Maasai rebellion in order to justify seizure of their land.[96] Following Harvey's lead, MacDonald threatened to raise the matter in the House, and did so in a debate on the colonies in July. 'Since 1904,' he told MPs, 'we have committed this great crime' against 'a peaceful people'. He called the Olonana deathbed tale 'much more like a Sunday School story than anything else', pointing out that land speculation in the highlands had rocketed since news of the impending move had emerged. A farm bought from the government for £55 at the beginning of the year had recently sold for £500. Harcourt wanted to know whose farm it was, but MacDonald could or would not say. He promised to ask his informant in BEA – 'a most reliable person'. Harcourt insisted he was not moving the Maasai 'on the strength of a dying speech'. Olonana's successor and other 'chiefs' had agreed to it: 'The Masai had come to a unanimous and even enthusiastic decision to move to the Southern Reserve.'[97]

Death on the Mau?

The move began in early June 1911. It was quite a sight, as Charles Miller has described: 'Napoleon's retreat from Moscow may have been a dress parade by comparison. Planned routes were forgotten as 10,000 Maasai, 175,000 cattle and over one million sheep sprawled out across the Rift and its two escarpments like nails spilled from a giant's keg'.[98] However, the official record says that the Maasai followed the four prescribed routes south. Moreover, they went quietly. ADC Harry Popplewell wrote: 'They appeared to be moving with absolute willing-ness and no pressure of any kind was put upon them.' By the end of July, all had reportedly left Laikipia except for Masikonde, Ole Gilisho, the 'Nyeri Masai' (the Dalalekutuk section) and some sick herds.[99] Days later, the cavalcade was in disarray as it reached the Mau in mid-August. Three of the four routes converged on the summit, where cold and heavy rain greeted the travellers. Grass was scarce, and the pathways became a muddy morass. One posse seems to have halted and turned back on those following, after learning there was insufficient grazing ahead. The result was chaos. Large numbers of Maasai began breaking back into the Rift Valley, flooding settler farms between Njoro and Gilgil, and the move was officially suspended. Popplewell wrote:

> Of the four roads used one passed round the South side of Naivasha Lake and crossed the Mau behind the Endabibi plains into Engattit. The other three passed respectively South and North of Lake Elmenteita and North of Lake Nakuru, reaching the foot of the Mau at three different points and converged onto one point at the sum-mit, whence the only outlet into the Reserve was one narrow track through the forest, taking at least two days to pass and practically without grazing. On the first route there was no difficulty; the Masai

passed right into the Reserve and stayed there. Those who used the other three routes became congested at the top of the Mau ... .[100]

Heavy rains and cold then added to their problems. The Maasai lacked adequate shelter, he said, and these conditions 'caused a certain number of deaths' of both people and livestock. The road became 'almost impassable'. Officers in charge of the move tried to push them forward, but the Maasai refused. Two-thirds of the total migrant population eventually returned to Laikipia, although at the time Girouard claimed that half the 'northern' Maasai had moved into the Southern Reserve. His first full report to the Colonial Office differed from Popplewell's in key respects. The initial back-up on Mau was caused, he said, by Maasai halting on the plateau because they found the grazing good. They put up their 'kraals' and decided not to move down to the plains. When the rains came the place became congested, and the roads out unusable. Citing reports by Archibald MacDonald (Director of Agriculture) and settler Dr Arthur Atkinson, he said the Maasai had 'suffered no great hardships ... and their losses are not of a serious character'.[101]

News of the alleged deaths soon leaked out. Humanitarians in Britain were alerted to what was going on via the Anti-Slavery and Aborigines' Protection Society. Three impassioned letters arrived at its London offices between the autumn of 1911 and January 1912.[102] The anonymous author, who signed him or herself variously 'A friend of the native' or 'An Englishman', described the plight of 'the unfortunate Masai of British East Africa who were being forced from their homes for a second time in spite of the statements of Mr Winston Churchill regarding the inviolability of Native Reserves'. Travers Buxton, the Society's secretary, presumed the handwriting to be that of a lady. The writer begged the Society to come to the aid of these unhappy Africans. The second letter, dated 25 September 1911 and simply sourced 'London', gave a purportedly eye-witness account of dead and dying Maasai and their stock on the Mau. Photographs of skeletons scattered in the bush, enclosed in a third letter dated 7 January 1912, appeared to confirm this report:

These people were driven from their homes without the slightest attempt being made to find out first whether the country they could go to could keep them, and secondly whether there was any possible chance of their being able to get there; the sole idea being to get them out as quickly as possible, and by different routes they were forced on to the top of the Mau Escarpment, it being an impossibility for them to get any further, owing to the absence of grass for their stock. About the 13th August a large deputation of these wretched people came in to the nearest Government Station and reported that their stock were dying of starvation, and also their little children, and the old people, whose food is practically entirely milk, and they begged to be allowed to return to the nearest available grazing. The local officials, apparently realizing the gravity of the situation, recommended that they should be allowed to return. The reply to this can hardly be believed. It was practically to the effect that they might stop and die, or go on and die, and that there was plenty of grass on the German border ...

Now to those of you who may happen to be the proud possessors of happy and well-fed children, try and imagine this scene – 'A rolling plain surrounded by impenetrable forests 10,000 feet above the sea, the coarse grass stamped into a sea of mud and dotted about with the temporary homes of the evicted Masai. From the circles of these rude shelters comes the wailing of the starving and dying children whose unhappy mothers unable to get the milk from the dying cows are trying to feed them with the best meat they can get from the dead cattle and sheep. Without, the never ceasing moaning of the hungry cattle is only broken by the screaming of the kites as they wheel overhead, a bitter wind and driving rain and sleet, while to complete the picture, on the scrubby and windswept cedars sit the ill-omened forms of the gorged vultures. As the short twilight rapidly deepens into night and the feathered scavengers slowly sail away to their resting places for the night, echoing up the little valleys comes the horrible cry of the hyena.

It was a case of 'death' where they were, certain 'death' in front, and perhaps what they feared most, death and outrage at the hands of the savage native soldiers at the orders presumably of those at the head of affairs. ... Now, though we are told that 'Vengeance belongs to the Lord', it is perhaps hardly too much to expect that the British nation will demand and see that 'Justice' is administered impartially, and ... any attempt to further carry out such atrocities shall be stopped at once and for ever.

The Society sent the photographs and first letter to the Secretary of State for the Colonies on 22 January 1912, and circulated the entire correspondence to various MPs including Harvey, its Honorary Secretary in the Commons. Thanks to Leys, the story of this injustice was not new to Harvey. He had been expecting something of the sort for months. With the claims of mayhem on the Mau, Girouard's nightmare had begun.

It is unlikely that Leys was the author of the anonymous letters – the handwriting differs, so does the overblown style, and such a detailed eye-witness account seems more likely to have come from an official involved with the move, or even from a local missionary. Wylie calls the writer of this last letter 'an ill-informed humanitarian', but one cannot assume this.[103] Writing to Harvey and MacDonald on 27 August, Leys said of the magistrate at Nakuru: 'He is the man who reported the affair originally.' It could be that this person decided to contact a pressure group in Britain, or passed information to a friend to disseminate. At any rate, Leys took up the same refrain and sought to investigate the facts. In October he wrote to Harvey, MacDonald and Murray:

> When, about the middle of September, over 30 of the Masai elders came to complain to the magistrate at Nakuru of the loss of life during the march, the doctor at Nakuru was sent up to enquire. He went from village to village asking in each the numbers who had died in the preceding month. He calculated that from two to four per cent of the population in the different villages had died. He saw no dead bodies. When he met with the Masai they had come down from the mountain forests and bush. Unfortunately he did not take names of the people who were alleged to have died in the mountains.[104]

The doctor, Henry Boedeker, had established the cause of death as exposure and famine, and claimed 'enormous' numbers of stock had also perished. But this finding was at odds with an official inquiry by the Director of Agriculture and Dr Atkinson, a close friend of Delamere's and his former travelling companion. Both these persons, wrote Leys with heavy sarcasm, were 'strong advocates of the move long before the CO heard of its having been proposed'. They reported seeing the corpse of one young woman, who looked too well nourished to have died of starvation, and a few animal carcasses; they did not think the livestock losses abnormal compared with those suffered by any European farmer. However, Goldfinch (who was to become involved in the resumed move as an Officer in Charge) told them that a 'large number' of cattle and sheep had died, and 'some old men and women and young children had died of exposure and starvation'. MacDonald and Atkinson conceded that there may have been some fatalities among the very old and very young about which they were unable to obtain direct evidence. Most of the people they had met looked healthy, and the children appeared particularly well nourished. In a separate report their companion, veterinary officer Richard Edmundson, said he had inspected Maasai livestock on the Elmenteita Plains and the Mau and found negligible losses.[105]

Leys later wrote that he did not believe this testimony for a moment, since he knew the inquiry had been conducted at breakneck speed in difficult bush country. MacDonald and Atkinson had been sent up to the Mau from Nairobi by train. They rode from the railway to the site of the alleged deaths, a distance of 30 or 40 miles. 'They went there and back in three days, on horseback. They dated their report on the third day from leaving the railway. To examine forest and bush country in the time was impossible. Their report, which states that there was no loss of life, is of no value. At the same time the doctor's estimate is probably much too high.'[106] Leys based his despatch on information supplied by the Nakuru magistrate, to whom he had written asking for facts and figures after a tip-off from an unexpected source – the Principal Judge of the High Court, Sir Robert Hamilton, who would hear the Maasai Case in 1913. Leys had written excitedly on 22 August of a conversation held that night with Hamilton:

> The Principal Judge of the High Court took me aside to tell me that he had learned that the Masai were finding grass and water so insufficient in the Southern Reserve that numbers of their cattle and sheep were dying. The milk supply is so short (Masai eat no farinaceous food) that the official in charge of the move is feeding some of the women and children on tinned milk. The streams are all polluted by thousands of dead cattle and sheep. Some of the Masai are trying to break back to the old reserve. But the grass is all trampled and eaten bare on the roads. And between them and their old country is 100 miles or so of settlers' country. The police are heading them back. <u>There</u> [his underline] is the danger. If a Masai is shot or a policeman speared a futile but bloody 'rising' is sure to follow. Some of the settlers have always prayed for it ... You can imagine the rest. I want to warn you that what has always been possible [a Maasai rising] seems now to be quite near ... I need not warn you not to let Judge Hamilton's name appear.[107]

Leys urged his parliamentary contacts to push for an independent inquiry, while lamenting that it would never take place and 'the real truth will never be known'. Indeed, the Protectorate administration saw no need for further inquiries, blaming the Maasai for their predicament. Sandford later wrote, repeating what Girouard had said in 1911:

> This check was undoubtedly the result of the Masai having proceeded too quickly. They pressed forward even when asked by European

farmers (through whose farms they were passing) to hold back, the farmers being desirous of trying to open up a trade with them. No police measures of any kind were adopted, testifying to the voluntary character of the move. It appears that the Masai did not suffer to any extent by the journey, and the losses reported were those which would naturally take place in large movements of stock.[108]

Eye-witness Thomas Ole Mootian remembered things rather differently. He was there, as a small boy, driving sheep and cows. A former colonial office clerk who became a politician and farmer, Ole Mootian was in his mid- to late-90s when interviewed shortly before his death in 1997, at his home at Olokurto on the Mau.

> We were pushed by force, by a white man called Bilownee [E. D. Browne] because he was the migration officer, accompanied by *askaris* [police/ soldiers] from Zambia, and others from Germany and others from Sudan. So now they were pushing us by force – it was not a joke. The *askaris* were holding guns. They were beating the people. When you stopped, they hit you with the butt of the gun. And if women made a joke or became lazy, they were caned. And when the sheep or cows became weak, they were killed ... Villages were coming, about 20 at a time. This is the movement, and *askaris* were on all sides, here and behind ... .[109]

Sandford's account remains the official version of events. It ignored the covert coercion (which the Colonial Office suspected), the lack of adequate planning (which surprised experienced administrators like Ainsworth, called in by Girouard to sort out the mess) and the timing of the move in atrocious weather conditions. And, of course, it said nothing about Maasai being driven at gunpoint to their Promised Land, and dying along the way. But in the margins of history, scribbled on a telegram from Girouard to the Colonial Office, is a sharper assessment by Harcourt: 'His suggestion that the move has "been pressed too quickly" is a piece of monstrous impertinence coming from him!' Girouard, whose coded telegram had ended with the words 'Masai are querbogen', meaning 'all quiet', was soon to lose his job for one impertinence too many.[110]

3
In Search of the Truth

> For heaven's sake don't think me a hero. I am only a cranky
> anachronism – a democrat in a country where every social,
> political and economic circumstance makes for slavery ... .
>
> Norman Leys[1]

It is impossible to prove whether there were any, and if so how many,
human casualties of the second move, besides the deaths from old age
and other natural causes that might be expected during any large-scale
migration. Yet there is an enduring conviction in the Maasai community that hundreds if not thousands of people died from starvation,
disease, exposure or gunshot wounds – a claim which suits the current
mood for reparations. This is refuted in oral testimony from elders who
took part in the move as children. Also, there are apparently no written
reports of casualty figures or descriptions of Maasai being deliberately
shot dead by British forces or African *askaris* under their control. This
chapter will begin by comparing different accounts of what happened,
then and now, before returning to the story of Leys's investigations and
the resumption of the second move. It will end by speculating about the
identities and motivation of other pro-Maasai whistle-blowers who may
have supplied Leys with information.

Besides Harry Popplewell, officers involved in the move reported some
evidence of sickness and stock losses, but no human deaths. Lionel
Talbot-Smith wrote of the contingent he escorted back to the Northern
Reserve after the move was aborted: 'There appeared to be a fair amount
of sickness among the natives, several requesting permission to rest
on the unoccupied farms, etc. before proceeding further. I gather that
fever and chest, possibly pneumonia, were the chief complaints.' He
noted 'considerable mortality among the sheep. In fact it would be no

exaggeration to state that the track was lined with their carcasses.' The Maasai told him the deaths were due to exposure on the Mau.² Crewe-Read made a tour of the so-called Promised Land (Mau Narok) and Il-Melili and said he saw no human remains and only negligible evidence that stock had suffered. He suggested that the sudden return of Maasai from the Mau to European farms in the Rift, accompanied by reports of heavy human and livestock casualties, simply amounted to an attempt to 'bluff the Government into returning them to Laikipia for ever'. He warned the elders he met that any return would be temporary, and that police were stationed all along the foot of the Mau to prevent those already in the Southern Reserve from following them north. Some elders told him they had no desire to return to Laikipia, and were surprised that others had turned back. They included Ol le Yeli (Ole Yiaile), an age-mate of Ole Gilisho's, who said he had had 'no difficulty whatsoever' *en route*, and did not wish to go back to Laikipia because he had found superior grazing.³

Curiously, human deaths were not mentioned in the 1913 plaint, though they were in the original affidavits; reparation was only sought for stock that had allegedly died as a result of the move.⁴ Leys himself developed doubts about the numbers of deaths, urging Harvey: 'Whatever action you may think wise to take in view of the facts, I strongly advise that no notice should be taken of the alleged deaths suffered by the Masai during the abortive move. The truth is not known. It probably never will be ... The Masai throw their dead into the bush for birds of prey. Up there, in the forest, 10,000 feet high, these are scarce. But in these two months some of the bodies will have gone and others be hard to find ... .'⁵ Lewis Harcourt was to say something similar, for different reasons, when put on the spot in the Commons in April 1912. Harvey asked him whether, 'in view of the very serious loss of life that occurred during the removal of part of the Masai south', he would give an undertaking that those still remaining in the north would not be moved until parliament had debated the matter. Harcourt said it had already been discussed in the House, in a debate on the colonies in July 1911, and there was nothing more to add. Harvey retorted that the loss of life had occurred since then; had Harcourt 'not seen a photograph showing skeletons on the roadside along the route which these unfortunate Masai took in moving south?' The response was cold:

> The loss of life was considerably less than that which usually occurs when the Masai move from one part of the country to another. I have seen the photograph. It shows only the skeleton of one woman, and,

as the honourable member must be aware, the Masai do not bury their dead, and it is therefore easy to discover the skeletons of those who die on the road.[6]

Maasai accounts

The Maasai believe it is unlucky to count people or cattle, or to name the dead. The majority of elders said in interview that they did not know who or how many (if any) had died while moving, and 'could not cheat' by giving false information. Daudi Ole Teka recalled that sick people were simply left behind: 'When somebody became weak when we were moving, we used to leave them in the camp. We used to leave the weak, somebody who was dying. Even weak cows, we left them behind. Many cows died. But I can't remember how many people died, because actually we had no problem like a disease that killed them all. So I can't cheat, I can't tell you the people who died or not.'

One of my oldest informants, Karanja Ole Koisikir, said: 'The white men ... chased us out [of Entorror]. They were not shooting us.' This was borne out by his neighbour, Sondo Ole Sadera. Too young to have been there, he was told about the move by his father: 'They were not using guns. They never shot anybody. They were just chasing them.' Among the eye witnesses was Nenaiduya Ene Olesuya, who moved south together with Ole Gilisho: 'We were forced to move, although nobody was killed. Some people died along the way from disease, and cattle too.'[7] Kitinti Ole Sadera corroborated these accounts: 'The soldiers were escorting us with firearms, though they did not use them. The only problem we faced while moving was starvation.' No one said that guns were fired at people; they were only used to put down sick animals, and to threaten the Maasai. Ntooto Ene Siololo remembered: 'The whites shot into the bush with guns, scaring the warriors. That is the work of the whites!' The very presence of white men was threatening. When Ole Teka was asked if the British used guns, he cried: 'Ay, ay, ay! You know, in those days, if a Maasai put on a coat, boys would run away in the field, thinking he was a white man. We feared the British. We never even wanted to see them.'

Maasai women bore the brunt of the move in many ways, since they had to construct temporary shelters, carry babies and household goods, care for the very young, the sick and the elderly, feed everyone, and give birth in very stressful circumstances. Women's narratives are understandably much more concerned with practical difficulties rather than the political events surrounding the move, since women were not party to these events or, at that time, to political discussion in general. Some may retort that the Maasai were nomadic anyway, and hence the hardship they faced while moving was nothing unusual. But forced migration bears no relation to the voluntary kind. Kirapusho Ene Gilisho, wife of Leperes Ole Gilisho and daughter-in-law of Parsaloi, was told about the move by her mother and other female relatives:

> Women were the ones who took care of the children during the move. They had [herbal] medicine in case a child became sick on the way. And when a woman gave birth on the move, they had to spend two or three days in one particular place in order for her to recover her strength. She would be given sheep's oil, blood and meat and after a few days she would be ready to move with the others ... There was not enough food, many people died of hunger on the way.

People drank a lot of soup, which could be carried in calabashes. Meat was stored in small boxes or gourds, which were also easily portable. Women also traded with the Kikuyu for food. When asked whether any women tried to resist the British, she said: 'Not at all. They had nothing to say to the white people. Women were trying to run away with their children, so they didn't have that in mind. That is always the work of women – when it comes to a fight, they run away with their children.'

Pantiya Ene Njapit, who claimed to be 100, took part in the move as a 'very big girl, more than 10 years old'. Again, she had not heard that anyone had died while moving. She remembered women working hard to build shelters: 'Women went and cut materials in the bush and made small shelters [to live in] for maybe one day or one week. If a woman gave birth, they stayed for one week.' Having settled in the south, she declared she 'really liked this place [Ngatet] very much, because this is where I grew up.' However, 'Maasai women did not want to be moved from Entorror, but no woman resisted the whites. They didn't have those powers to resist. They had no say, because women in Maasailand have no say. They just follow what the men decide.'

At 75, Kurito Ene Sengeny was too young to have been involved in the move, but stories were passed down to her. Her family settled initially in Loita, and 'chased' the Loitai away. 'I heard that when people were coming [south] they encountered a lot of problems, like people becoming sick on the way, other people dying from diseases, or maybe killed by wild animals.' Women gathered medicinal roots and plants to boil and give to the sick. As for food, the migrants lived on milk, blood and meat, and bartered for other goods. Kiter Ene Lemein was about 90, and at the time of interview lived near Ololulunga. Her family had moved first to the Mara. 'When they were moving,' she said, 'men were taking the cows and women were coming with the calves and children. Women were treating sick people using herbal medicines like *o-sokonoi* and *e-sumeita*. Entorror was better than this place because there were no diseases for animals and people. When they went to Mara, people were getting colds and malaria.' This was a constant refrain that will be fully explored in Chapter 5.

Suspension of the move

To return to 1911, after news emerged in BEA and Britain of the chaotic situation on the Mau, the move was suspended, the Secretary of State issuing orders that those Maasai who had not been accommodated in the south must be allowed to return to their old reserve for the time being. Some were temporarily allowed on to European farms in the Rift, including 150,000 acres of East Africa Syndicate land at Gilgil. They would not be moved again until Girouard had satisfied the Colonial Office that there was sufficient water in the south, and that the move could be properly supervised.

Leys wrote to Harvey: 'Nobody here really knows what happened. The Government has issued a statement to the press, not only minimising last month's trouble, but attributing the blame to the Masai – accusing them of being in a hurry to go South! ... I greatly fear that the doctor in Nakuru – a weak man – has had pressure put on him to withdraw his report or change it, and also that the magistrate there painted the picture so luridly as to give the white washers a chance. The feeling of the country is very disturbed.' This description of the magistrate's account tallies with the highly emotive anonymous letter to the Anti-Slavery Society quoted in Chapter 2, almost suggesting that the town magistrate, Ronald Donald, could have written it.[8]

Meanwhile Girouard sent Harcourt a curious review of the Maasai affair, in which he tried to deflect attention from his political actions to Maasai sexual mores. In a 16-page policy review, he suggested that someone should be placed in charge 'who I feel certain could produce definite results by moral suasion ... Personally I have been and am still a sincere well-wisher of the Masai. I fully realise however that the time has come for definite and stricter measures to be adopted with them. This will be kinder in the end ... strict administration based on schemes of development and improvement in their sexual relationships, must tend to their uplifting and to their ultimate good.'[9] Girouard is likely to have taken his moral cue from missionaries whose views he had canvassed on the advisability of placing the Maasai in one reserve. Bishop Allgeyer of the Catholic Missionary Society had told him: 'You are only too well aware of the moral condition of this people, and I feel sure that our mission will have a much better opportunity of improving this people if they are once brought together and firmly controlled by Government.' Allgeyer later warned of the danger of spreading cattle disease because the Maasai moved about so freely, before reiterating the moral argument: 'The Masai is a crafty enterprising fellow, not wanting in intelligence. He is however a coward and will easily yield to a strong hand. He is not inaccessible to serious moral training and betterment, if properly taken in hand ... .' Bishop Peel also wrote of their 'low moral condition' and 'sexual excesses' which would lead to ruin unless they were 'saved' by the missions.[10]

Girouard did, however, make some valid points in this despatch. Splitting the Maasai in two had been 'the primary mistake' made by government. Anyone could have anticipated that trouble would result from allowing mass movements of stock, at the time of circumcision ceremonies, to pass from one reserve to the other over European farms. On Laikipia the Maasai were 'hemmed in all sides' with no room to expand. Then, in a thinly disguised attack on the officers in charge, he said he could find no evidence of any coherent Maasai policy being applied in the north or south between 1904 and 1909, the year he had arrived in BEA. Neither had the government taken seriously its duty to administer the 'northern' Maasai. It was the circumcision ceremonies of 1910 that first drew his attention to the situation, he said; it was then that Olonana sought him out and asked for his people to be united in one place. The move had brought one important fact to light: the number of stock owned by the 'northern' Maasai had been greatly underestimated and was now set at around 200,000 cattle and two million sheep. The Northern Reserve was clearly far too small for Maasai needs. A strong new policy was needed, to include the abolition of warriorhood, the replacement of hut tax with a stock tax, and the encouragement of cattle trading. As for the stalled move, Girouard emphasised that 'rumours of large numbers of people and cattle dying on the Mau are entirely without foundation'. The most astounding statement was his denial that he had seized land for European settlement:

This is not a question of the seizure of native lands but of changing the pasture lands of a nomad pastoral tribe from one part of the Protectorate to another in order that they may be brought together ... and under one Government with a view to retrieving gross errors made in the past.

The Colonial Office was not fooled. Senior civil servant F. G. A. Butler minuted on the despatch that Girouard appeared to be 'in a hurry to cover up what looks like a disastrous experiment'. London repeated its instruction that the Maasai were not to be coerced in any way. Meanwhile, Girouard appointed John Ainsworth (co-architect with Hobley of the 1904 Agreement) to take charge temporarily of the Maasai and their suspended move. Leys applauded, calling him 'the best administrator in the country'. Ainsworth was summoned from Nyanza Province to Naivasha. After a quick reconnaissance of one area, Ainsworth reckoned that the move could not be carried out in one go, but should be spread over several months and done in relays. He expressed surprise at the lack of planning and preparation. Later, in his memoirs, he also doubted that his old friend Olonana had ever supported the move: 'At the time I was somewhat surprised to hear of this request and was inclined to suspect that outside pressure had been brought to bear on the Laibon.' He recommended that the government should immediately set up a station in the Southern Reserve, start water conservation schemes, make a proper survey of the reserve and obtain expert reports on its grazing capacity. On a lighter note, his account of the rescue mission included a ditty penned by one of the officers involved in the move, who pleaded to be given any other work (*kazi*) but this:

Oh dear, will it never be finished for ever?
This Promised Land exodus, how it does pall!
There's Ol-Beress in the deuce of a mess;
He can't keep his cattle at all.
The settlers abuse us, there's none to excuse us.
Their furrows are broken, their crops trampled lie.
Oh! powers that be, any *kazi* give me
Except further tramps with the giddy Masai.

Ainsworth returned to his old post in Kisumu after telling the 'northern' Maasai that the move was merely postponed, and that they would be expected to obey orders later. Incidentally, when news of the impending Maasai legal action emerged, Ainsworth was effectively accused by Lord Delamere of having instigated it, after Delamere passed on a rumour he had heard from J. K. Hill (of the East Africa Syndicate) to the Chief Secretary, Charles Bowring. Ainsworth later wrote: 'I took extreme exception to the rumour that I had instigated the Masai to take legal action. The fact, however, that the Masai did take such action was not in my opinion anything very terrible. If they really thought that they were being unjustly treated it was a much more civilised way of doing things than going into rebellion.'[11]

Surveys of the reserve

It had become clear that the western extension of the Southern Reserve was already occupied by other Maasai, principally the Loitai section, and was more or less fully stocked. One idea was to extend it again, to take in Trans-Mara – land to the west of the Mara river, which the Maasai call Olorukoti. This had not been on offer under the 1911 Agreement; Harcourt told Girouard he must inspect it thoroughly first. The surveys that followed were more revealing of their authors' interests than the territory in question. Archie MacDonald, Director of Agriculture, led the first tour of inspection and reported back on 11 November. He was accompanied by J. K. Hill, a settler called Chaplin, DC Rupert Hemsted and McGregor Ross, in his capacity as Director of Public Works. They were told to check the adequacy of water supplies and the stock-holding capacity of an area measuring roughly one million acres. Before setting out, Ross wrote home: 'Although I don't think much of my companions [with the exception of Hemsted, "quite a pal of mine"], I can't help having a very interesting tour.'[12]

MacDonald reported that the area was 'practically unoccupied', contained 'some of the finest grazing land in the Protectorate and is exceedingly well watered'. Total cattle carrying capacity was put at 185,000 head. As to why it was unoccupied, except by the Siria Maasai, this was explained away by smallpox and rinderpest having wiped out much of the population. There were said to be no signs of tsetse fly. At the Colonial Office, Butler remained sceptical, minuting that this was 'all very glowing'. He urged caution until the other reports came in, and they learned what the Maasai thought of the area and whether or not they gave their 'full and free assent' to the resumed move. Harcourt agreed they must wait: 'But I do not conceive that we are asking the Northern Masai for assent to the <u>renewal</u> of the move. They assented originally and the move is only postponed owing to unforeseen events. If they refused now we should be in an impossible position.' (His underlining.) He queried the 'rosy account' of the water supply, asking if this was intended to save spending the £2000 already set aside to improve it? Butler replied that if they were talking about any other governor but

Girouard, the response to this would be a resounding 'no'. As usual, CO minutes on the correspondence give the best indication of what was passing through the official mind, and the state of the relationship between London and its African outposts. (By 28 November, an exasperated Harcourt was scribbling: 'I give it up; I can't waste the remnants of my brain of his [Girouard's] riddles [*sic*].'[13])

Hill's report was even more effusive, prompting Harcourt to comment: 'Mr Hill seems to have inadvertently dropped into paradise. It should be remembered that he is Manager to the E.A. Syndicate on whose land many of the N. Masai are now located and that the E.A. Synd. Property adjoins the Northern Masai Reserve which is to be vacated!'[14] Hill put the carrying capacity higher than MacDonald, and saw no signs of fly or other biting insects though these were said to infest the lower reaches of the Gori river. He thought the Siria cattle were in superb condition, and only cautioned that the water supplies in some areas would not allow thousands of head of Maasai cattle or sheep to sup simultaneously. Hill was 'sad' that settlers would see such beautiful land passing into the hands of 'natives' – it was the finest and best watered country in the whole Protectorate. Both Hill's and MacDonald's remarks about tsetse fly contradicted those of other observers, and what the Maasai – and those officials most closely in contact with them – knew to be true. Browne, for example, had written to Pickford about Maasai avoidance of the Mara area 'on account of the fly'. Ross seems to have told Leys, who then told Harvey, that the proposed extension 'was far too good for the Masai but that there must be a curse on the country. What he meant was that for generations the area had been uninhabited and that no one knew the reason. The explanation may be sleeping sickness ... part of it is in the "fly area".'[15]

A second inspection of Trans-Mara was made in November, and led by Hemsted. Maasai representatives joined him, and allegedly said that the country was suitable. Ole Gilisho picked four men to join a third trip, again led by Hemsted in November. This time, there was a change of mood. 'They returned nothing but evasive answers,' wrote Sandford, 'at the same time admitting that the grazing was sufficient and the water supply accurate.'[16] Hemsted described the thoroughness of the inspection: 'They appeared to be keenly interested, inspecting all the herds of cattle met with, and making careful inquiries as to the permanency of the streams, the nature of the grazing, etc.' However, they would not commit themselves in any way. Hemsted thought their remarks 'generally rather foolish' – they claimed their cattle would fall into game pits and their people would catch sleeping sickness from the Kavirondo.

They insisted on consulting Ole Gilisho before saying any more. Hemsted suspected that Ole Gilisho had told them 'to adopt this non-committal and rather unassailable attitude ... I have every reason to believe they are perfectly satisfied with Ol-orukoti '[Trans-Mara] as a cattle country ... .'[17]

The same month, Leys claimed to have proof that Maasai leaders had been ordered to sign a ready-made petition, 'purporting to be a voluntary request to go to the Southern Reserve. They were pressed to sign on several occasions before they finally submitted and agreed. There is no use asking if that was how it was done. No Government admits to such practices.' Ramsay MacDonald sent a copy of the letter to Harcourt, without the signature, saying it was from his most reliable correspondent.[18] Informant Ole Mootian confirmed that such a petition had existed, and was not signed willingly: 'There was nobody who agreed to be on the list. But if you have a gun pointed at you, will you agree or won't you?'

Collyer admonished the Laikipia returnees at a meeting in November at Gilgil, which Ole Gilisho attended. He told them in no uncertain terms that the present 'impasse' was largely their own fault: 'They had always refused to be of any assistance to the progress of the country by trading their cattle, and had insisted on sitting still and increasing their livestock in spite of repeated warnings.' The government was disappointed that such an 'intelligent tribe' was content to sit still and allow communities like the Kikuyu and Kavirondo to leave them far behind in development.[19] But behind Collyer's public bluff lurked private doubts, which he later shared with J. W. T. (John) McClellan, PC Naivasha: 'What I can't get over is that the Government have got themselves into a position when they can give no orders. They have no one to blame for this but themselves. I wish I could see some way out of the impasse; it has been going on too long.' When British representatives met Maasai leaders the following February at Naivasha, several Maasai who had moved south called on the 'northern' Maasai to join them. Olonana's cousin Ngaroya (*sic*, Nkaroyia), regent to the young prophet Seggi in the Southern Reserve, said it was Seggi's wish that all the Maasai should now move south. Ole Kotikosh for the Purko and Ole Seti for the Keekonyokie also reportedly said that members of their sections who had not already moved wanted to be allowed to do so. But Ole Gilisho and Masikonde held firm, saying they definitely did not want to move. Ole Gilisho was said to have prevented several Purko who wanted to move from doing so.

This was borne out by Collyer, now posted to the Southern Reserve. He sent a message with nine of Ole Kotikosh's warriors to the ADC

Rumuruti on 11 April, saying they were coming back to fetch their property; he hoped this official, Mogorr, would insist on the old men handing over the livestock. He had heard through his headman, whose wife was Maasai, that Ole Gilisho was using threats and curses to stop anyone moving south; some individuals wanted to. Maasai were apparently claiming, 'We have defeated the Government, they can't throw us out of Laikipia, Laikipia is ours, let anyone who has relations or property go back there.' Soon he was telling McClellan of his feelings of utter powerlessness:

> The leader of this movement is one Ol le Kisando, and I understand that a number of Laikotikosh's people are going to try to break back ... They may be on their way now ... Ol le Yeli thought they would try and cross the railway at night close to Elmenteita. The bulk of old men who promised to go Westward now seem to be sitting tight on E Uaso Nyiro again, but what is the good when I can't enforce my orders; it is only making a fool of myself and doing no good; if I was allowed to break kraals or something of that sort I might do some good.[20]

The continued defiance of the returnees became supremely embarrassing to the government as the months passed. It made nonsense of the official claim that the Maasai had asked to move in the first place. Leys wrote to Harvey of the 'abominable truth of the Masai business – a far worse affair than I had thought'. He had learned that Girouard had intimidated the Maasai into 'asking' to be moved, and that Ole Gilisho had been threatened with flogging and deportation by Lane unless he persuaded his people to petition the government to move them south. By April 1912 Leys had left Mombasa for a new posting at Fort Hall, about 80 miles east of the Northern Reserve. He said he hoped 'to take a journey, if I can find a pretext, to get an interview with Legalishu' and other Maasai. 'My idea is that if they tell a straightforward consistent story which if true convicts government agents of using threats of punishment for expressing their real opinions, I shall be able to write an official letter demanding an enquiry.' There is no further mention in the Harvey Letters of any such meeting. (Leys later stated he had never written or directly spoken to any Maasai leaders.[21]) By 23 May, Ole Gilisho was being moved south again. The crunch had come at a British–Maasai summit on 21 May at Naivasha, when the order to move was reiterated. This time, there was no opposition; even Ole Gilisho appeared to have changed his mind, agreeing to move south with the first batch and to

help people settle once they had crossed the Mau. As the physical distance widened between them, Leys was no closer to getting firsthand
evidence from the *ol-aiguenani.*

The Colonial Office had given the go-ahead for the move to resume
after Girouard assured Harcourt that the reserve was ready to receive the
Maasai. But their meeting in May 1912, at which this order was given,
was a stormy one that paved the way for Girouard's departure. Harcourt
grilled the Governor about his denial the previous autumn that he had
promised settlers land on Laikipia before the Maasai had vacated it. In
October 1911 it had been decided that no such promises were to be
made until the question of the Maasai move was settled. At the time,
Harcourt had doubted if Girouard was telling the truth. Now his worst
suspicions were confirmed.[22] For his part, Girouard saw no inconsis-
tency in his actions.

Ole Gilisho, meanwhile, was playing a waiting game – news of his
impending legal action was about to break. On 28 June, Scottish lawyer
Alexander Morrison told the Colonial Office that he, A. D. Home and
A. W. Buckland had been instructed to act for Ole Gilisho and others in
legal proceedings against the government. Cashmore is probably correct
in suggesting that Ole Gilisho volunteered to start moving when he did
in order that he could meet Morrison outside the reserves, in relative
freedom from state control.[23]

The Cole case

While all these intrigues were unfolding, the Cole affair was scandalising
the country. There are parallels between this and the Maasai story, as
Leys, MacDonald and others pointed out at the time. Settler Galbraith
Cole was Lord Delamere's brother-in-law, and a son of the Earl of
Enniskillen, who had arrived in the Protectorate in 1903 and taken up
land in 1905. He farmed sheep at Kekopey (adjoining Delamere's
Soysambu ranch) in the Rift Valley, and on Laikipia. He was deported to
Britain in the autumn of 1911 after an all-white jury acquitted him of
the murder of a suspected sheep thief on his farm, shot as he ran away.
Reports differ as to whether the victim was Maasai or Kikuyu, though
Cole's son Arthur told me he was Maasai.[24] Cole had set out to find some
Africans he believed had stolen a sheep, came across several men skin-
ning a sheep a few miles from his house, assumed they were the thieves
and opened fire when two of the men ran off. The fatal bullet went clean
through the victim's back and out through his stomach, leaving half his
intestines spilled on the ground. Wrote McGregor Ross: 'The dirty points

about Cole's action were that he shot the native and left him to die ... and that he did not report the case but lay low. It was only found out as a result of native report 12 days later.'[25]

Once apprehended, Cole did not deny the murder charge. Far from it: he told the jury that he shot to kill. The jury was only out for five minutes, and ignored Judge Hamilton's direction to return a guilty verdict. The acquittal caused outrage in certain quarters, not least at the Colonial Office, which demanded that Cole should be deported for 'conducting himself so as to be dangerous to peace and good order in East Africa'.[26] It was feared that African unrest and European demonstrations could follow the acquittal, so Cole was effectively being accused post-trial of stirring racial tension. Settlers petitioned government to withdraw the order. Cole's own counsel, Sir Edward Carson, advised him to defy it and stay in East Africa. Harcourt became increasingly angered by what he saw as Girouard's failure to follow orders and publicly condemn the actions of both Cole and jury. When he learned from *The Times* of 9 September that the Governor was suggesting the deportation order had been issued by the Secretary of State, he wired to say the order must be issued in the Governor's name alone. He accused Girouard of being uncooperative and trying to shirk responsibility.[27]

These events had the Protectorate buzzing with speculation. Lively coverage of the case in the British and East African press raised public awareness of another alleged outrage of empire, and laid bare the attitudes of some settlers – never mind the media and other Europeans – to Africans. Leys was again involved in igniting interest at home, using Harvey and MacDonald as conduits. (Twenty-four questions were asked in the House about the Cole case and Maasai move between May 1911 and August 1912, mostly by these two MPs. Questions solely about the Maasai were raised 18 times in 1911–13.) Leys probably heard the inside story of the murder from his colleague Dr Boedeker, who was sent to examine the body. In letters to Harvey, he said that in comparison to the 'hopeless' case of the Masai: 'I am really more hopeful about the useful-ness of the Cole Case, and I do hope you are getting papers published – the official reports I mean. Even yet the Governor has said nothing publicly about the Case. No one has. For all the public knows he is on the side of the men who shoot natives when they run away ... Once again a thousand thanks.'[28]

In a letter home, McGregor Ross confided: 'I am staying with Leys, who has ceased to be excited re the Masai move since it is a *fait accompli*. He is still on the warpath about the Cole case, however.'[29] MacDonald linked the two issues in parliamentary debates, while in BEA *The Leader* newspaper went further in reporting and then scotching a rumour that the real reason for Cole's deportation was because he had advised the Maasai.[30] There is no evidence for this whatsoever; the Cole family as a whole was hardly in favour of Maasai land rights. Galbraith had to give up land at Ndaragua, east of Thomson's Falls on Laikipia, when the Northern Reserve was created; in exchange, he was given land in the Rift. After the second move, he got more land on Laikipia.

Girouard kept telling the CO, much to their annoyance, that rampant stock theft had driven Cole to commit the crime. Settlers said it was the bane of their lives, and if the government could not protect them, they would take the law into their own hands. Leys did not sympathise with the murderer, but he doubted the wisdom of the deportation:

> What happens to Cole is nothing. The real problem is what is to happen to prevent assaults, trivial or murderous, and other abuse of political and social superiority done by the 4000 Europeans in the country to the 4,000,000 natives of it. You can't deport every European who gives a boy 20 lashes for breaking machinery by his carelessness, or a man who smashes a boy's head with a bottle for impertinence. Deportation is not a method of dealing with crime. It is only a way of disgracing an occasional criminal.[31]

Lord Delamere cabled the *Daily Mirror* to protest at the deportation. Both national and provincial British newspapers ran the story, with strong leaders for and against the verdict. For example, the *Manchester Daily Guardian* castigated settlers and their apologists:

> Some of the white colonists in British East Africa badly need to be taught that they are subjects, not kings. Their attitude in this matter is that of lordly slave-owners in the Southern Sates of America resenting the control of a 'Yankee' Government. Cole is one of a number of settlers, mainly drawn from the wealthier classes in England, who seem to have taken out to East Africa ideas alien to those associated with British rule over subject-races. We cannot permit a class whose deadening domination is only now being lifted from the English countryside to practise a coarser despotism on the natives of Africa.[32]

Leys suggested that some good could come of the Cole affair if right-minded people were 'driven by it to the roots of the difficulty … The only complete remedy of course depends on economic, social, religious changes' as well as an overhaul of the justice system. But the case sent out dangerous signals to Africans, because it could 'destroy the confidence we fondly imagine natives to have in our system ... .'[33] Leys seemed to be spoiling for an African rebellion, which did not materialise in the expected form. Its absence confirmed his often repeated view of 'native docility', combined with a *Heart of Darkness* dread of the enemy within. Professor George Shepperson, in his introduction to the fourth edition of *Kenya*, notes: 'Norman Leys was obsessed with the fear of African insurrections.' In this, the socialist anti-imperialist had 'something, however slender, in common with the sentiments of the declared imperialists of his day ... with their dread of the consequences of the breakdown of law and order and Pax Britannica'.[34] Leys urged white reform, not black revolution, while acknowledging the likely necessity of the latter. He had at the start of this correspondence told Harvey to be ready for native rebellions in ten years' time. After the Maasai returned to Laikipia, he had halved the waiting time: 'The two best judges I know believe that a Masai "rising" is probable in the next five years.' On several occasions he had told Harvey that some settlers would relish this, for they believed the Maasai were long overdue for a punitive expedition – something they had taken part in many times, but never been subjected to themselves.[35]

Cole did not stay away long. He returned to BEA soon after World War I began, and picked up farming where he had left off. He was apparently allowed to return after his mother wrote to the CO to say he was wanted for the war effort in East Africa. This was unlikely, since he was virtually crippled by arthritis. Leys, by this time banished to Nyasaland, could not resist asking Harvey: 'Do you think it possible that I might share in the amnesty? Or does the Colonial Office think my offence worse than Cole's murder?'[36]

Girouard soon followed Cole out of the country, initially on home leave. By July, he had resigned and gone for good. It wasn't simply a case of having deceived the CO about the Maasai move; he had also lied to Harcourt about promising settlers farms on Laikipia. The row over this raged long after Girouard had left. One settler – Russell Bowker – even demanded that the government should pay him rent, as Girouard had verbally pledged, for accommodating the flocks of settlers displaced from the Southern Uasio Nyiro to make way for the 'northern' Maasai.[37] Girouard was also in trouble with the CO for a host of other reasons. He had begun to build the Thika tramway, effectively a private railway line for a handful of estate-owners who needed to get their produce to market, without official blessing and having lied about what it would cost. He had enraged Harcourt by telling the press that Cole was only being deported on the orders of the CO. He was far too close to the settler community for the CO's liking, and it only knew half the story. But there were also personal reasons to resign. He had been offered a lucrative new job, a directorship with the shipbuilding and armaments firm of Armstrong, Whitworth & Co. near Newcastle upon Tyne. His wife was sick and had been advised not to return to East Africa. Privately, his marriage was on the rocks and would soon be over; Gwen had petitioned for divorce on the grounds of desertion and adultery. As Governor, the stigma of this was unthinkable.[38] He was tired of the problems facing him, tortured by insomnia, and certainly prepared to go.

Build-up to the Maasai Case

Ole Gilisho and his supporters continued to appear to go along with the move, and cooperate with government, while preparing their forthcoming action. Ole Gilisho was acting with three fellow Purko – Parmuat, Enessering and Engeness (also known as Parmuat Ol le Sopin/Sopia, Narsering Ol le Kool and Engenes Ol le Guriangei, all likely to be misspellings). In Civil Case No. 91 of 1912, the plaintiffs claimed that the 1911 Agreement was not binding on them and other Maasai of Laikipia, and that the 1904 Agreement remained in force. They claimed £5000 damages for stock lost during the earlier, aborted move south and sought an injunction against McClellan, PC Naivasha, and Hemsted, Officer in Charge Southern Reserve, to stop them preventing the return of any Maasai and their stock to Laikipia or compelling the same to move from Laikipia. They granted power of attorney to Messrs. Bischoff & Co. of London.

According to an affidavit sworn in the High Court, Stephano Ole Nongop – a young Mombasa-based Christian temporarily employed by the legal firm as 'boy and interpreter' – was sent to Ole Gilisho's camp on 14 and 15 June to deliver legal papers and a letter.[39] Morrison was staying nearby with farmer A. S. Flemmer, and asked Stephano to make the first overture. The whole affidavit smacks of intimidation, and is indicative of the witch-hunt that had now been launched against people with Maasai sympathies. Stephano, who was studying at Buxton High School in Mombasa, said his master, the Reverend George Wright of Mombasa, had 'lent' him to Morrison for the trip to Nakuru. On 15 June, Stephano had read the papers over to Ole Gilisho, and he and other Maasai told Stephano they were all being moved against their will 'and that but for fear of the European they would remain in Laikipia … Not one said he was moving of his own choice.' They said they had all agreed to employ advocates in BEA and London, and to file a suit in order to achieve their rights. They had agreed to incur costs worth 2000 bulls or £5000, but they needed to be able to collect them from other Maasai – 'at present they had no opportunity, being hurried continually on without a moment's leisure.' Ole Gilisho instructed Stephano to write two letters to Morrison, agreeing to pay the necessary fees.

Next day, said Stephano, Ole Gilisho told Morrison he was 'very, very glad' that he had agreed to act for them. Cost was no object. They arranged for Ole Gilisho to come to Flemmer's farm on 17 June to sign the papers and give more detailed instructions, but this was forbidden by Crewe-Read (by now ADC in Charge of the West Route, Maasai move). He told the Maasai leader he 'must go with the safari'. He also appears to have refused to give Ole Gilisho time to cut at least 40 bullocks out of his herd, which he wanted to leave with Flemmer for future sale. Morrison then developed a fever and was unable to do any work. It is unclear what happened next. But this was the start of a process of gross interference by the local administration (no such orders came from London) in communications between plaintiffs and lawyers, which was clearly designed to stall or derail the legal action.

An unnamed friend of Leys allegedly first suggested to Ole Gilisho that he should see a lawyer. 'I procured the best one in the country for him', Leys told Harvey. This was more than Leys ever admitted in print, although in letters to the Governor in July 1912 and February 1913, forwarded to the CO, he confessed to having engaged Morrison and arranged for him to visit Ole Gilisho. Morrison was 'an old and intimate friend'. Leys later elaborated: 'There were two of us concerned in getting a lawyer for Legalishu. They have sent for the other man to go to Nairobi to be cross-examined and browbeaten. He is a simple-minded person and they may intimidate him into resigning. I wish they had sent for me.'[40] On learning that he had seen a lawyer, Hollis (now Secretary for Native Affairs) pressed Ole Gilisho to make a statement explaining his actions:

Statement by Legalishu, Enderit River, 21st June 1912
Legalishu stated that two Europeans came to him at Pusi Lokony on Soisambu [Lord Delamere's ranch] on the evening of the 18th instant and offered, on payment of 40 bullocks, to assist him in regaining Laikipia. On being asked by Mr Hollis if he had called these Europeans, he replied: 'Is it possible for a black man to call a white man?' He then stated definitely that he did not call the Europeans, but that they came to him. He had never seen them before.[41]

Ole Gilisho reportedly said he wished to accept the offer, because 'if a man offers something you like, is it likely you would refuse?' He asked the government not to 'lower his prestige with his people', saying he realised there was an order to move and he was now prepared to coop-erate to the best of his ability. The statement was read back to him and he agreed that these were his words. But two days later Morrison sent the plaint to G. H. Goldfinch, an Officer in Charge of the move, to be witnessed. By 25 June it was declared a civil case in the High Court at Mombasa. Leys explained Ole Gilisho's behaviour before Hollis as the tactics of a man under pressure. He had already told Harvey of how the *ol-aiguenani* had been threatened with flogging and deportation: 'Probably Legalishu, being terrified out of his wits, did say he was ready to go and repudiated my friend the lawyer. He will probably play a double game for some time.'[42]

Charles Bowring (Acting Commissioner February to October 1912) forwarded to London a letter of protest from Leys, in which Leys admit-ted arranging Morrison's visit to Ole Gilisho. Bowring hoped that Leys would be removed from the colonial service without delay. He played down the threat to government of the pending action:

I am advised by the Attorney General that legally the position of the Government is perfectly secure, and the incident is apparently being engineered either by unscrupulous persons who wish to trade on the credulity of the Masai to their own pecuniary advantage or by so-called 'sympathisers' with the tribe who, like Dr Leys, are obsessed with the idea that the Government has adopted a policy of systematic and continual oppression where the interests of the natives are con-cerned, despite the fact that officials, non-officials and missionaries who possess a far longer and more intimate acquaintance with the indigenous population hold an entirely contrary opinion.[43]

Morrison denied having 'touted for employment by the Masai'. When his authorisation to act for them was officially questioned, Morrison retorted:

From the beginning of the year, it has been well known that Legalishu wished to have a lawyer to help him and his people, and wanted to pay a yearly retainer of 100 bulls. So far as I know Dr Leys had nothing to do with this, and the suggestion came from another and high-placed official. Dr Leys ascertained from me that I was will-ing to act for the Masai on terms and, I understand, recommended

me. The Masai also approached Mr Buckland ... It is not the case that
I sent to Legalishu before he asked me to come to him.

The point may seem academic now, but it was to cost Leys his job. By
the end of 1912, Ole Gilisho was talking of sending a Maasai deputation
to London to appeal directly to the Secretary of State and MPs. Nothing
seems to have come of this.[44]

Stephano's affidavit also provides a rare insight into how relatively
educated and urbanised Maasai, living some distance away, viewed the
predicament of those on Laikipia. He was one of a small group of such
people, all young and mission educated, some of whom had been influenced by the American missionary John Stauffacher, who assisted their
upcountry kin to press the case. Stephano told the court: 'Before leaving
Mombasa Mr Morrison asked me why the Masai were leaving Laikipia,
and I at once told him it was for fear of the Government. I said so as all
I had heard of the movement was to that effect, and it is obvious the
Masai themselves would not voluntarily consent to leave Laikipia. I said
this before I knew Mr Morrison had any business with the Masai, and it
is the absolute truth.'

The most important member of this loose group was Molonket Ole
Sempele. A Keekonyokie Maasai, he played a peripheral but key role in
the Maasai resistance, and is chiefly remembered today for his church
and political work. He broke away from the Africa Inland Mission (AIM)
in 1930, over the issue of female circumcision, set up with others an
independent church and school, and became involved in the Kikuyu
Central Association and the Maasai Association, together with Ole
Mootian. As a boy, he was befriended by the newly arrived Stauffacher
who wanted help in learning the Maa language. Their rather intense
friendship was to last for 40 years, until Stauffacher finally left Kenya.[45]
In 1904, he helped Stauffacher to try and convert Maasai living in the
Rift Valley. When the AIM established a mission on Rumuruti after the
first move, Ole Sempele joined it despite great opposition from his family and tribal leaders. This is where his early schooling seems to have
taken place. Kenneth King describes how Ole Sempele then sold some
cattle to pay for his passage to the United States in 1909, to seek higher
education – the first Kenyan African to do so there. Funded by the mission, he attended an all-black college in the South, and in three years his
political consciousness was transformed – largely because his experience
of southern-style racial prejudice gave him new insights into the situation in Kenya. On return home in 1912, Ole Sempele found his people
in dire straits and set about helping them to resist the second move.

(Morrison, however, said he returned from the United States on hearing of this.[46]) Stauffacher, who had tried to get the Maasai to oppose the first move, had in Ole Sempele's absence been urging another young Maasai, Taki Ole Kindi, to persuade the elders to resist. Most felt that armed resistance would be unrealistic. Ole Sempele teamed up with those Maasai who favoured legal action, contacted Morrison at Mombasa and offered his services, but the AIM stopped him from acting as an interpreter and advisor to the plaintiffs.[47]

Colonial Office records, and elderly Maasai who knew him, tell the rest of the story. On 13 September 1912, Lee. H. Downing, Acting Director of the Kijabe Mission, wrote an indignant letter to the Judge of the High Court on learning that 'a Masai boy named Mulungit is, in [sic] behalf of his people, about to enter a suit against the Administration'. Downing expressed great surprise, since Ole Sempele had been with the mission eight years and not said anything about this. The mission had given him permission about five weeks earlier to attend a meeting in Nakuru, with a European from Mombasa and some Maasai. He was away a long time, and on learning from Ole Sempele that Morrison was trying to get a reserve pass for him, Downing wired Ole Sempele with orders to return. According to Downing, Morrison had 'influenced' Sempele to take action against the government, but the young man now 'felt very sorry' and wanted the missionary to stop 'the attorneys doing anything more in his name. With this request I comply most cheerfully.' The mission did not remotely 'approve of one of our boys championing such a cause'.

Morrison's colleague Home had perhaps acted naïvely in sending Downing a letter, via Ole Sempele, in which he outlined the Maasai action and asked Downing to help Sempele get into the reserve without a pass. If he was arrested, wrote Home, 'please at once wire us'. He revealed that Ole Sempele had applied to the High Court for an order enabling him to start an action against government officials, and compelling them to give him a reserve pass. As he saw it, Morrison thought there was a straight choice:

1 Either the boys in the reserves are absolute slaves, and as such must obey each and every arbitrary order of the PC and not dare to approach His Majesty's courts; or
2 they are free men and as such have a constitutional right of approaching the judgement seat with their petitions. The Court must decide.

Downing strongly disagreed; it was up to the mission to decide what its 'boys' did. He accused Morrison of 'using Mulungit without consulting us, when he is subject to our direction'. He claimed Mulungit now saw the 'folly' of his actions. Soon afterwards, Ole Sempele signed a statement before the Attorney-General and Acting Chief Secretary Monson in which he swore he had only agreed to see Morrison because he thought he was a 'Government man'. Downing had told him not to have anything more to do with the lawyer. The Maasai plaintiffs and their legal team had lost a potentially valuable, and literate, intermediary.[48]

One of my key informants, the late Reverend Peter Kuyoni Ole Kasura of Ilmashariani Africa Inland Church, near Narok, was born and brought up in the same settlement as Ole Gilisho – Enjoro-Emotoroki, or River of Ducks, near Lemek. He joined the mission at Siapei in 1922, and Ole Sempele was among his teachers there. Born in Entorror, he took part in the second move as a small child and was probably in his late 80s at the time of interview in 1997, but was still working as a minister. He had vivid memories of both Ole Gilisho and Ole Sempele, and understood from personal experience how the latter was pulled in two directions. Ole Gilisho was so determined not to lose Ole Kasura to the mission, that he reportedly chased after him in a plane belonging to the Narok DC. If he had not joined the church, said Ole Kasura, he would have become an age-set spokesman; he had already been chosen. Ole Sempele had been in the same position. He gave up his traditional status for a very lowly role with the AIM, before joining a separatist church movement.

'This one called Molonket,' said Ole Kasura, 'was converted by the missionaries when they were in Entorror, and they took Molonket overseas and he went learning there.' As for Ole Sempele's view of the loss of Entorror, 'he told me that he hated the way the British took their good land and forced them to a bad side. But he never wanted the Maasai to fight with the British'.

Obstruction of the lawsuit

The British attempted to block the forthcoming action in several ways, according to Leys and the lawyers. The Maasai were prevented from selling cattle to raise the legal fees by a ban on cattle trading and through placing the reserve in continuous quarantine.[49] Intimidation of Ole Gilisho continued. Governor Belfield ordered Home and Morrison that all consultations between them and their clients should take place up-country as far as possible, rather than Mombasa. Morrison saw this as an attempt to swell the legal fees, and hence stop the Maasai from pursuing the action. The lawyers were subsequently refused passes to enter the reserve, and their clients were prevented from leaving it. Entry passes were also refused to other known sympathisers, notably Ole Sempele, to stop them acting as interpreters and intermediaries. This isolated the plaintiffs in the reserve, and prevented the interpreters from making contact. The High Court refused to allow the payment of advance court fees to be postponed (it is not clear why these were demanded in advance), while preventing Ole Sempele from going into the reserve to fetch and sell his livestock, thus forcing him to organise a whip-round among impoverished urban Maasai. Morrison told the Colonial Office: 'If my clients were criminals in jail I should have readier access to them than I have at present.' Since Ole Gilisho was illiterate, it was imperative that passes should be given to literate intermediaries: '[He] has no one among his following able to read. If I am to send a confidential letter to Legalishu it is necessary I should also send a confidential person to read him the letter.'[50] The CO ordered the Governor not to obstruct the lawyer or his clients, or make unnecessary difficulties. It was primarily concerned about public opinion.

But reserve passes for two young men known only as Peter and Juma (described respectively as Swahili and Maasai) were repeatedly refused during August and September 1912. Simultaneously, Ole Gilisho and his fellow plaintiffs were also refused permission to leave the reserve; Hollis told Morrison the 'political situation' did not allow it. When Home applied for a reserve pass, in order to collect evidence and arrange payment of court fees, he was told this was not possible because the government did not know of any 'legal action in the Courts which would render it necessary for you to meet clients in that locality [the Southern Reserve]'. After Harcourt intervened, passes were finally granted in December to Andrea Murioki and Karumba Masai to enter the reserve to obtain signatures and evidence of stock deaths. However, Harcourt's own office was not exactly helpful: when Morrison tried to see the Secretary of State in person, he was rebuffed by senior civil servants Herbert Read and W. C. Bottomley. Read noted: 'We said as little as we civilly could and got rid of him as soon as possible'.[51]

Making sense of Leys and his circle

> With you there is daylight. Here, helpless fools are being throttled in the dark.[52]

Leys expected to be sacked quite early on, as soon as Girouard had got wind of his activities (as opposed to his attitudes, which he did not attempt to disguise). He wrote to Harvey on Christmas Eve 1911: 'The Governor has sent a message to me to the effect that he hears I have been conspiring against him and wishes to know about it. I replied that face to face and divested of our official relations I will gladly tell him what I think and what I have done.' But he was not directly challenged at first, and held on. By 22 July 1912 he had confessed to Acting Governor Bowring that he had arranged Morrison's visit to Ole Gilisho. Writing of 'our mad firebrand Leys [being] in rather deep waters now', Ross described how Leys had 'written a final counterblast to Sir Percy ... he says it is none of the "timid slush" that he wrote to Sir P. two years ago' and he 'expects to get the sack for it but that he will refuse to go! Talk about stormy petrels!' He stayed with Leys in September, and they sat up talking about the issue until after midnight. The local administration had by now asked the CO to dismiss Leys, and he knew it: 'He has got his tail up pretty stiff, but Mrs L is rather upset about things.' Bowring ordered Leys to 'refrain from any further interference, either directly or indirectly, in political matters outside the scope of your official duties'. Leys eventually paid dearly, going into what he called 'exile' in Nyasaland in spring 1913, after a period of home leave. But he shrugged it off, telling Harvey he should 'entirely ignore my exile ... I have paid very cheaply for my opinions, less, I am sure than thousands pay daily in England. It is sheer nonsense to consider me in any but the most ordinary light.' He told Harvey not to call him a martyr – he did not have 'a single drop of martyr's blood'.[53]

What was his motivation in taking up the whole issue? 'I don't care a rap for the Masai, particularly,' he wrote in September 1912. It went further and deeper than that. He once told Gilbert Murray that he had been looking for a social injustice to expose since his earliest days in Africa. With the Maasai, in the short term he wanted to bring about a judicial inquiry and expose imperial policy: 'I would like to see the history of the whole affair used to illuminate the methods and aims of the administration of our African Protectorates.' He was also determined to oppose and expose settler hegemony – 'They are demanding and receiving a position of privilege which includes the real control of the country and their own enrichment by lavish grants of land, literally worth millions of pounds.' In the long term, as he explained to Harvey in his first letter, he wanted his correspondents to help forge a new colonial policy. That entailed challenging the profit motive that he believed drove policy, and led to the oppression of Africans. 'I have yet to see an African government that is not mainly directed by consideration for the men, whether in England or in Africa, who profit by African land and African labour … I want you and your kind to lead and change policy when the chance comes. I am tired of inferior imitations of Kipling's heroes.'[54]

He believed 'we are laying the foundations of an evil system that generations to come will struggle with,' and asked Harvey to help dismantle it. Local dissenters like himself were powerless to make real change, he felt. He also knew he was inclined to get things out of proportion. He believed East Africa needed a different type of governor – one who was prepared to challenge settlers and their ilk: 'I long for a man to prove to them [the settlers] that their rights over Africans and Africa are no greater than an English farmer's over other Englishmen. In Africa slavery is dead but liberty is not alive. In Africa I want men from home who have courage to oppose local interested parties … .'[55]

The rhetoric suggests that he saw himself following in the footsteps of the great anti-slavery campaigners, and people like Morel. In the same breath, however, Leys warned Harvey not to take any action over the Maasai or to think unkindly of Girouard. This ambivalence of tone runs throughout the correspondence, signifying the mixed loyalties of the servant who lives by the system he scorns. In the very next paragraph Leys urged his friend: 'I hope you will leave the Masai question alone,' for he believed the Secretary of State would do his best, and he did not want Girouard to go: 'He has brains. But he is vain, absurdly fond of popularity, dearly loves the rich and titled. And he bluffs … But if you only knew how rare and precious brains are in Africa you would sympathise with my wish to keep him.' By the late 1920s, Leys had undergone a fundamental change of heart. In a letter to the Anti-Slavery Society, he declared: 'I am gradually coming round to the view that it will prove impossible to carry out even the most moderate reforms in Kenya so long as the European Colony is in existence.'[56]

What was the quality and reliability of his sources? Collyer was clearly one of his main informants and not, by all accounts, a person given to exaggeration or politicking. This is confirmed by his family, who believe his early death was hastened by his bitterness over the Maasai move and his cavalier treatment by government. He was not the kind of person likely to blow the whistle himself, or to be openly disloyal, but he had a strong sense of fair play. His great-niece, Veronica Bellers, comments:

I think it is unlikely that he wrote anonymously; he wasn't a disloyal man. When Leys says Arthur was sickened and embittered, I think the point is that he sat alone night after night with this awful dilemma. He was being told what to do by Girouard but in his heart he didn't want to do this. At one point Girouard rather forces Arthur to say things he didn't want to say. I feel, in being made to say he was a friend of the Maasai, and what government was suggesting should be done was in their interests, was *not* what Arthur believed. He believed tribal land as it was demarcated should be as sacrosanct as freehold land in Britain. This was the thing that tore Arthur apart, being made to say things he did not want to.[57]

Judge Hamilton was also Leys's confidante, and the magistrate at Nakuru had at Leys's request supplied information about the crisis on the Mau in autumn 1911. In May 1912, Leys told Harvey that 'a new correspondent' had contacted him, 'an officer now engaged in helping to move the Masai, begging me to get it stopped'. In his next letter, he said this brought to two the number of officers involved in the move who were corresponding with him – '[they] warn me to look out for resistance and its consequences'.[58] (It is unclear whether he included Collyer here, but he was no longer involved with the move.) To speculate, these men are very likely to have been Goldfinch, possibly also McClellan and/or Browne. George Hammond Goldfinch, an assistant game warden since 1907, assisted the Maasai in returning to Laikipia in the autumn of 1911, and became Officer in Charge Mau Camp during the resumed move. He reported the original deaths on the Mau in 1911 to Archie MacDonald and Atkinson, as mentioned earlier. On the evidence of his letters to David Davies MP (the gist of which was communicated to Harcourt without naming their author), and his later correspondence with the Anti-Slavery Society, he also reported them to a wider circle.

Goldfinch had come to BEA from South Africa in 1904, entered government service in 1906 as a stock inspector, and was an assistant game warden for Nakuru District from 1907 to 1923. In World War I, he was on the intelligence staff. He had been a Master of Foxhounds (MFH) in England, and on arrival in East Africa became second Master of the Masara Hunt from 1904–08. He owned a small pack of home-bred hounds, and also ran a stock farm near Nanyuki. First, to describe what Davies told Harcourt.[59] In September 1911, Davies had recently returned from a visit to the Protectorate. Writing on 28 September, he told Harcourt that he had during this trip 'had an opportunity of seeing something of the Masai'. Since coming home, he had received 'several very urgent letters from an Englishman whom I met at Nairobi', on the subject of the Maasai move. He enclosed extracts, saying he believed they provided genuine grounds for an enquiry. Davies's informant wished to remain anonymous 'on account of the acute feeling which exists in the Protectorate against those people who dare to criticise the native policy of the Government'. But he could vouch that the man had no axe to grind, was totally honest, and simply 'cherishes a deep concern for the welfare of the natives'. The clues to Goldfinch's identity are that the informant was said to be 'a good sportsman', and Davies had 'lived with him in the wilds for two months'.

The extracts themselves are puzzling. They refer to a third, unnamed party – 'one of my very best friends who has helped me with my information ... one of the finest sportsmen who ever holloa'd to a hound'. Either the letters were written by Goldfinch, pretending to have got information from a fictitious friend, or they were written by someone else to whom Goldfinch had supplied information. They told a now familiar story: how speculators had wanted to get hold of Laikipia and turn the Maasai out, how the Maasai had supposedly asked to move, and their 'chief' had been threatened with jail if they did not budge. The writer pointed out how law-abiding the Maasai had always been, and therefore did not deserve this harsh treatment. But he warned: 'If their cattle die of starvation they can hardly be expected to do anything except rise.' Ole Gilisho's appointment as regent to Seggi was described as a bribe he was now 'most bitterly regretting ... [It] was a clever move on the part of the Government, as they apparently said to him "Now we are going to do you a great honour and make you a Regent of Lenana's heir, paying you a good salary, but you will have to go and live in the Southern Reserve, and you will be a very big man there over the Masai".' It was revealed that Girouard, on being told of the crisis on Mau by the DC Naivasha, wired back: 'Move them to Tsavo.' The letter writer commented: 'This was a most dreadful thing to say' because Tsavo was 'a "wait-a-bit" thorn desert, infected with all the known forms of Tsetse fly, except I believe one form. It is also a most poisonously malarial spot.' Such an order amounted to a death sentence:

> I consider it is the most disgraceful thing that has occurred under the British flag during the last century, and I cannot tell you how sore and disgusted I feel about it ... It makes one feel absolutely ill to hear of such doings against a luckless native tribe and a law abiding tribe too.

Goldfinch (if indeed he was responsible for these revelations) showed his hand openly in his later letters to the Anti-Slavery Society. He kept it informed, through the mid-1920s, on such hot subjects as the contested 'Delamere exchange' of land on Laikipia (enclosing copies of his protest letters to the *East African Standard*), forced labour on coffee plantations, outbreaks of plague in reserves, Samburu land claims north of Laikipia, the Maasai generally, the plight of the 'Dorobo' and the impending move of the Momonyot from the Loldaika Hills, east of Laikipia, to the Southern Reserve.[60] In the Momonyot case, he took thumbprints and statements from a 'Dorobo' and so-called head of the Momonyot, Guaisain or Guaisaiu ol Legeshaur (*sic*, the writing is illegible). In remarks echoing those of Purko already in the south, Legeshaur feared they would lose most if not all of their stock to disease and starvation if forced to move, and begged to be allowed to stay put.

Goldfinch agreed with his assessment, and in a cover note to the Society added: 'This old man is rather fussed and comes and worries me and I have told him that I can do nothing for him here but that I will send his petition home to people who are really interested in people like himself.' In other letters, there are references to 'my friend McGregor Ross'; clearly, he was part of the same circle. He told the Society that the Native Affairs Department was 'a disgrace', Maxwell the Native Affairs Commissioner 'a knave and a fool ... who has persistently robbed and persecuted the natives'. He said he was not anti-settler (he referred elsewhere to both Cole brothers as friends) but simply thought 'that the native should have a fair deal'. Finally, an eight-page handwritten statement headed 'The Masai Scandal' made totally plain where his sympathies lay: 'no tribe of savages have ever had to put up with what they have in the last few years.' As to what the Society made of Goldfinch, it told T. Johnston MP: 'G. H. Goldfinch is a very rough diamond, but he does know what he is talking about.'[61]

Goldfinch and McClellan were the two officers who, in the summer of 1912 while Ole Gilisho was moving, had handed him legal papers from Morrison about granting power of attorney to a London law firm – not assistance that British officials were obliged to give. Goldfinch was commended by DC Talbot-Smith for his extreme care in assisting Maasai to return to Laikipia, in words that conveyed his conscientiousness and concern: 'the success of the removal of the Masai back to their old Reserve from around Nakuru is due to the trouble and care Mr. Goldfinch took, in explaining the road to the natives, and assisting them in every way in his power.'[62] As for McClellan, he was a confidante of Collyer's and shared his attitude towards Africans. While Acting Secretary for Native Affairs in 1908, during Hollis's absence on leave, he had argued for the sympathetic treatment of Africans and support for chiefs, and railed against unnecessary jail sentences and pass laws. He had also denounced the suggestion that Africans should be given small reserves in order to force them out to work as 'wholly wrong and indefensible'.

Goldfinch later told Travers Buxton of his 35-year friendship with McClellan, who was by 1924 also corresponding with the Society. He added: 'you could not possibly have a better authority on the Masai and their troubles'.[63]

One might also speculate about E. D. (Edward) Browne, ADC Laikipia. His 'move reports' were fairly sympathetic to the Maasai, and he went on to become the first British administrator of Tanganyika Maasailand, where he and his close relationship with the Maasai became legendary. Tidrick mentions that he 'came down to Tanganyika convinced that the Kenya Masai had had a rotten deal [over the move] and determined to see that the Tanganyika Masai got a better one'.[64] Either way, the indications are that middle-ranking officials closely involved with these events were speaking with firsthand knowledge through Leys. And after all his doubts about the alleged human deaths incurred on the 1911 move, and his shillyshallying with Harvey over this, one of his sources finally reassured him of the truth of this central claim: 'I have just learned from a trustworthy source that the story of loss of life on the Masai move was true.'[65]

Overall, the evidence points strongly to Goldfinch being Leys's key informant all along, apart from Collyer. He was on the Mau at the crucial time in the autumn of 1911, and was in a position to have taken the photographs of Maasai skeletons that were sent anonymously to the Anti-Slavery Society. He may have passed information to the author of the letter that accompanied them, but the handwriting does not match his. He escorted Maasai groups back north and was involved in taking them south the following year. His work as a game warden brought him into intimate contact with Maasai and 'Dorobo' in reserves that partly overlapped with the northern and southern game reserves. He thoroughly knew and understood these two environments and the fine balance between humans, domestic stock and wild game, which could be upset so easily and disastrously. From the later Anti-Slavery correspondence it is apparent that he had a particular interest in questions of disease and resistance of both stock and humans moved into unfamiliar environments, and in these letters he made his attitudes to the alienation of African land, as well as other abuses, transparently clear. He may also, in supplying S. V. Cooke and the Society with information, have been instrumental in helping the Samburu to retain the Leroghi plateau, and certainly tried to stop the Momonyot removal by lobbying the Society which in turn used its parliamentary contacts. He also took up with Travers Buxton the case of Olonana's half-brother Senteu, who had been condemned to internal 'exile' in the Meru area in 1919 for his alleged complicity in the Purko–Loita raids of 1918–19. Having befriended Senteu, now sick, penniless and reduced to hanging out with 'Dorobo' in the forests, Goldfinch believed it a scandal that the old man could not at least be allowed to return to his own area to die in dignity. He was in fact allowed to return to Loita by 1925, which suggests that the Society may have intervened.[66]

Other anecdotes about Goldfinch reveal he was lamed for life in 1906 while hunting lion on horseback with game ranger Blayney Percival and a settler called Lucas. Goldfinch was mauled and bitten on the thigh, while Lucas died from his injuries a week later.[67] Percival's brother Philip described him as 'a dear old gentleman of independent means'. Elsewhere, Trzebinski mentions a prank involving Goldfinch at the Norfolk Hotel, Nairobi. He was sitting quietly having a beer, by a window overlooking the veranda, when a group of 'cowboys' (settlers on ponies) lassoed him out of his chair.[68] All in all, Goldfinch seems to have run with the hare and hunted with the hounds when it suited him, maintaining friendships across a wide social and racial spectrum. This gave him insights, and empathy with Africans, that many other Europeans in BEA lacked. Although he went public in attacking Delamere in letters to the local press, many of his settler friends probably had no idea of his clandestine pro-African activities, apart from Berkeley Cole whom he described as sharing his views on the Maasai.

Some like to party

Several of Goldfinch's colleagues among officers and other whites employed on the Maasai move saw it in very different terms to himself, as a happy and profitable time. Anglo-Irish immigrant Richard Gethin, employed as a junior stock inspector under Popplewell, wrote: 'The Masai Move was a very well run show … To everyone's regret [it] came to an end early in 1913.' When it was over, Hemsted threw a party at Narok for all those involved, presumably Europeans only. They were presented with letters of thanks from the Governor for doing such a good job in difficult circumstances. 'It was a most successful party and went on till daylight.'[69]

Why was it profitable? 'The Masai Move was a most pleasant period, the staff had plenty of big game shooting, everyone being well mounted, owning anything from three to four ponies each.' To a poor young man like Gethin (from a 'good' family but now in reduced circumstances), who had previously managed settler Edward Powys Cobb's farm at Loydien, Naivasha, and had recently returned to East Africa without a job, the chance offer of work supervising these safaris was very welcome. The salary was 300 rupees a month, but there was extra money to be made through selling Somali ponies and the skins of animals shot *en route*. By spring 1913, 'I was now well in funds from sales of Ponies, Lion and Leopard skins, and my pay, I had saved about £500 which in those days went quite a long way.' The pony-selling scam went like this: all Somali ponies, mules and donkeys coming from Abyssinia had to be quarantined for a month at Rumuruti before proceeding to Nairobi. Captain Charles Neave and Bill Kennedy, respectively Chief Stock Inspector and a veterinary officer employed on the move, vetted the ponies before letting them go. Officers with the move took their pick of the bunch, handed them over to their syces (grooms) for smartening up, and left Rumuruti with up to 20 ponies per safari. 'On arriving at Gilgil a wire would be sent to Nairobi giving the date of arrival at Naivasha, and people would come up by train and take their pick, paying about double the price we had paid for them. The ponies had by now been groomed, clipped, well fed and looked quite a different animal to what he or she was on arrival at Rumuruti.'

Unlike Goldfinch – whom he crossly described elsewhere, ticking him off for ivory trading – Gethin defended the necessity of the move, and rebuked the 'chairborn Politicians at home [who] thought we were being very hard on the Masai'. He described each safari as consisting of an unspecified number of Maasai families, around 10,000 head of cattle and 5000 sheep and goats, and covering about ten miles per day: 'There were permanent bomas for the stock but the Masai had to make their own arrangements for sleeping accommodation.' He was employed on the route that took in Gilgil, Elmenteita, Mau and the Lemek valley. Interestingly, he said there were no warriors in the parties he escorted: 'To make things more difficult, the Masai Moran refused to have anything to do with the Move, clearing off to the Southern Reserve on their own, leaving old men and boys to herd the stock during the move.'

In a surprising postscript, Gethin hinted at the good relations that had been forged between official escorts and Maasai. The day after the party, Popplewell suggested that they should 'ride over to where chiefs Masacondi and Legulesha [*sic*] were camped to say goodbye to them, as they had both been a great help to us in getting the safaris together at Rumuruti'. They ran into some trouble with a lion, however, and he did not describe any final goodbyes. Later, he referred to meeting 'a number of old friends who I had brought down from Rumuruti' while on safari in the Southern Reserve. They included Masikonde. The Maasai he met then were hungry, he said, and only too pleased to buy flour from him.

At the end of the day, no truth had been established – only the certainty that something smelled rotten about this entire affair. Leys had departed, as had his old adversary Girouard, his once promising career in the colonial service having foundered on the Maasai scandal. Collyer had probably taken several secrets to the grave, dying 'a deeply bitter man' according to his great-niece. Olonana's true intent can only be guessed at, on his deathbed and beforehand, but the record of his duplicitous role does not look good. The only winners appear to have been the European settlers who delightedly took up farms on Laikipia, and who had undoubtedly applied great pressure on local government to achieve this goal. Part II will describe the aftermath of these events, beginning with the 1913 lawsuit in which the Maasai attempted to regain part of their former northern territory.

Part II
The Aftermath

4
The Court Case

> It must be admitted that it often seems unreasonable to apply
> civilised law to simple savage life.
>
> Sir Charles Eliot, Commissioner of
British East Africa[1]

This chapter will describe the 1913 Maasai Case and appeal, and critiques
of and responses to the judgement. Shortly before the case came to
court, the Maasai won an injunction restraining the Crown from moving or continuing to move them from Laikipia. However, this proved
useless: it came through on 10 April 1913, a fortnight after the last
Maasai had left Laikipia. Ole Gilisho's son-in-law Murket Ole Nchoko,
misspelled Ol le Njogo by the British, became first plaintiff in Civil Case
No. 91 of 1912, which begat the Maasai Case.[2] He was described in the
plaint as a leading *moran* of the Purko section. His seven fellow plaintiffs
were Purko and Keekonyokie.

What had become of Ole Gilisho? As a signatory of the 1911
Agreement, who now stood accused with 17 other Maasai of having no
authority to enter into such a contract, he could not be both plaintiff
and defendant. One assumes that this was the main reason why he did
not, having initiated the action, take the case to court. He may also have
tired of the whole business after being subjected, over many months, to
threats and character assassination by British officials. One informant,
Karanja Ole Koisikir, said: 'Ole Gilisho did not go to court because he
had already realised that he was defeated – the court judges were whites.
Ole Nchoko went to court because of his stupidity.' Several informants
recalled Olonana's warnings 'from beyond the grave' that Ole Gilisho
would be in dire trouble if he pressed ahead. One of his sons, Shoriba,
claimed: 'Ole Gilisho dropped the case because he was told by the *ol-oiboni*

[Olonana] to leave it.' But Olonana had died two years earlier. One can interpret these stories in at least two ways. First, Olonana may have issued an earlier warning not to mess with the white people; he had prophesied that they would eventually leave the country, and the land they had taken would be returned. He may have warned Ole Gilisho in particular, since he threatened to jeopardise Olonana's position by resisting the move. Second, it may be wishful thinking to reinvent Olonana the collaborator as Olonana the protector, defending Maasai interests and 'saving' the much-loved Ole Gilisho.[3]

The plaintiffs claimed as individuals and also on behalf of the Maasai of Laikipia and the Maasai generally that the 1904 treaty was still in force and effect. The first defendant was the Attorney-General of BEA, sued as the government's representative. Defendants 2–19, all Maasai signatories to the 1911 Agreement, were said to have had no authority to enter into it, and it was therefore void except with regard to them. The plaintiffs claimed that the Maasai (with the exception of defendants 2–19) were still entitled to Laikipia. They demanded £5000 damages for failing to provide a connecting road between the reserves as agreed in 1904, unspecified damages for loss of stock that had died while moving, and for depreciation of stock wrongfully removed from Laikipia. To the best of the plaintiffs' belief, defendants 11–19 were in favour of their legal action.[4] Defendants 20 and 21 were J. W. T. McClellan, PC Naivasha, and Rupert Hemsted, Officer in Charge of the Southern Reserve.

The plaintiffs claimed that the 1911 Agreement was void for five reasons:

1   The plaintiffs and other Maasai had never consented to it or authorised the Maasai defendants or anyone else to agree to it on their behalf;
2   the defendants had no authority to alienate the interests of minors and unborn children of the Laikipia Maasai;
3   it did not benefit the Maasai generally, or those on Laikipia. The government was in a fiduciary position to the Maasai (that is, they were trustees, as a result of the 1904 Agreement and later declarations by the Secretary of State) and had thereby gained financially;
4   the Maasai had had no independent legal advice before signing it;
5   defendants 11–19 had not signed it voluntarily.

Most significantly, the warrior plaintiffs threw down a gauntlet to the older generation. They said the Maasai defendants, unless expressly authorised by members of the tribe, had no authority to 'deal with' land and no authority in this case because 'according to the ancient tribal custom of the Masai elders such as defendants Nos. 2 to 19 can give the advice only, but the actual decision in any particular case rests with a council of the Moran or warriors'.[5]

Ole Nchoko's original affidavit described his personal stock losses and claimed that many people had died during the move for a variety of reasons, though he gave no numbers.[6] He and other warriors on Laikipia had refused to move south when first ordered to do so, knowing that the reserve was 'utterly unsuited' to their needs for five reasons. These were: 'the total absence of water in most parts; the insufficiency of the water holes where there is water; the permanent presence of fell disease [it is not clear what this was] – noxious to both men and beast; the presence of the sleeping sickness and the tetse [*sic*] fly in great numbers; the unsuitability of the ground, over the whole area, for pasturing sheep; the presence of the Southern Masai who require the whole of the area for their cattle leaving no ground available for our people'. But they were told that if they did not move, the government 'would use the utmost force to compel us' and duly 'inundated the district with their armed askaries [*sic*] in charge of white officers'. Significantly, although the orders were issued through Collyer, the plaintiffs did not blame him: 'Throughout the whole of these proceedings Mr Collyer treated us very kindly and considerately and told us that it was not his order but the order of the Government and as such must be obeyed.'

Ole Nchoko claimed to have personally lost a quarter of his cattle and half his sheep while moving, to sickness that did not exist on Laikipia, as well as lack of food and water. Speaking of the 'northern' Maasai as a whole, he claimed that 'our cattle have died in hundreds and our sheep in thousands during such movement'. Many cattle had contracted pleuro-pneumonia and rinderpest, and been shot dead by government officers; no exact numbers were given. He complained about anthrax in the Mosuru and Mara areas of the Southern Reserve, fatal to both humans and cattle; 79 Maasai had recently died of it.[7] However, in the plaint that went before the court Ole Nchoko's losses had been scaled down. The total stock allegedly lost was 97,910 cows and 298,829 sheep, worth £200,000. Total depreciation of stock was approximately £100,000, while Laikipia was valued at £1 million.

Before turning to the case, I want to look briefly at Ole Nchoko, a member of the Il-Meiruturut or right-hand circumcision group of the Il-Tareto age-set. The written literature is unhelpful – he only appears in passing. I traced three of his sons, now living at Ololulunga near Narok, and also asked other informants about Ole Nchoko's involvement in the case. One son, Salaton Ole Nchoko, said: 'My father was acting as a

mediator between the Maasai and the white people, because he understood a little Kiswahili and English, so when the white people gave a
message, he took it and delivered it to the Maasai. In other words, he
was a translator.'[8] However, he never went to school, and could not read
or write English. He was a government-appointed 'chief', not a customary *ol-aiguenani* for the age-set. He was also an assistant 'chief' to Ole
Gilisho, and they were described as 'good friends'. Salaton did not know
why his father had gone to court instead of Ole Gilisho, or anything
more about the case. His two older brothers, Saiponyari and Olmengo
Ole Nchoko, also said they were not told much about it. Said Saiponyari:
'My father was actually a little bit informed, so that is why he went to
court in order to defend the land of the Maasai, simply because he knew
a lot of people, especially whites. He had a white friend who was actually
telling him how to go about it. He tried to go to court, but all in vain. He
was finally defeated because the government was for the whites.'

His work as a mediator and translator appears to have begun in
Laikipia, where he forged links with friendly Europeans. 'My father tried
to get a legal action because, first of all, he was a person who was
together with the white people before they were evicted from Entorror,'
said Olmengo. 'So now, when people were forced to come [move], those
friends of his were actually advising him on what to do. So that is why
he was the one who tried to go ahead [with the case] because he was
being told how to go about it. My father was somehow teaching the
Maasai how to take this legal action.' Neither son had heard about
lawyers being involved, or even about the two Agreements. 'All I know,'
said Olmengo of the post-1913 period, 'is that the white people were
always going with my father, telling him the words he was to tell the
Maasai, also telling him things he was going to tell Ole Gilisho, who was
the superior leader. White people were coming almost every day to his
place, talking to him and taking him to bigger towns, and he stayed
there for several days at government offices.'

Asked whether they felt any bitterness about the outcome of the court
case, Olmengo said: 'We are not bitter. We are even happy about it,
because my father acquired wealth as a chief for the colonial government
and that is why we are what we are now.'

The 1913 case

The Maasai Case was dismissed in the High Court at Mombasa on
26 May 1913, on the grounds that the plaintiffs' claims were not cog-
nisable in municipal courts.[9] The Maasai Agreements were ruled to be

not agreements but treaties, which were Acts of State. They could not, therefore, be challenged in a local court. It was impossible for the appellants to seek to enforce the provisions of a treaty – 'the paramount chief himself could not bring such an action, still less can his people'. Claims for damages against the first, twentieth and twenty-first defendants (the Attorney-General, McClellan and Hemsted) were also thrown out, for their actions in carrying out the terms of a treaty were ruled to be as much Acts of State as the treaty itself. The plaintiffs' arguments were not fully aired. The case was effectively dismissed on a technicality, a preliminary point of law. Judge Hamilton ended his judgement by quoting Lord Kingsdowne in the Privy Council case *Secretary of State for India* v. *K. B. Sahaba*, who said of actions by the Crown in violation of treaties: 'It may have been just or unjust, politic or impolitic, beneficial or injurious, taken as a whole, to those whose interests are affected. These are considerations into which their Lordships cannot enter. It is sufficient to say that even if a wrong has been done, it is a wrong for which no Municipal Court of Justice can afford a remedy.'

The judgement centred on the status of a protectorate, in which the King was said to exercise powers by virtue of the Foreign Jurisdiction Act of 1890. The Crown claimed that BEA was not actually British territory, and therefore the Maasai were not British subjects with any attendant rights of access to British law – 'But East Africa being a Protectorate in which the Crown has jurisdiction is in relation to the Crown a foreign country under its protection, and its native inhabitants are not subjects owing allegiance to the Crown but protected foreigners, who, in return for that protection, owe obedience.'[10]

One of the plaintiffs' central contentions was that the Maasai who signed the second Agreement did not represent them or the Maasai as a whole. But the Crown viewed the signatories as 'persons whom the Commissioner and Governor, acting on behalf of the Crown, *chose* as representatives of the Masai tribe with whom the Crown could enter into such agreements'. (My italics.) Alexander Morrison, lawyer for the Maasai, commented in a letter to Harvey in June: 'Sir Robert Finlay's opinion differs from that of the learned Judge. He has advised us that the Action lies in the Ordinary Court and that the question for decision is one of fact depending on the authority of the Chiefs to bind the tribe according to Maasai custom. I accordingly advised my Clients to appeal but I have not yet been able to see them.'[11]

On 21 June, the *Nation* newspaper published an anonymous article on the case, written by Harvey. (The Harvey Letters reveal its authorship.[12]) Entitled 'A Naboth's Vineyard – The Sequel', it was the follow-up to a feature Harvey had written two years earlier, which told the Maasai story to date. Harvey used heavy sarcasm and an appeal to Christian conscience:

> When the day comes for the historian to have access to all the records which are now out of reach even of the eyes of a Colonial Secretary, it will be possible for a remarkable story to be written of the way in which, beneath the shelter of a British Protectorate, the hands of Sir Having Greedy have been stretched out to seize the possessions of a savage tribe, unhappy in their too great wealth ... Will the Masai appeal from this decision to the Privy Council, and what will the outcome be? The imperfect story of our dealings with this people is not pleasant reading, but at least we can be glad that under British rule it should be possible for a subject tribe to impugn the justice of the action even of the highest of the King's officials, and to appeal ... to the law behind them all, and the great tribunal which embodies it.[13]

Harvey reminded readers that the possibility of a reserve in the south had 'strangely enough' been discussed by Eliot in 1903. In the White Paper on the Maasai, Deputy Commissioner Jackson had said of this proposal: 'Let those who advocate the Kedong Valley and to the south of it visit the country in the dry weather. No sane European would accept a free gift of 500,000 acres in such a place. Why, then, try to force such a place on the Maasai? Higher ground, and a considerable area of it, is absolutely necessary, and it is impossible to deny that the Maasai are entitled to it.'[14] As for how the Maasai now felt, Harvey went on: 'the desire of these uneducated folk to keep what was once their own had taken an unconscionable time in dying.' Leys told Harvey that he hoped the article would 'prepare the public for what is coming'. He feared the worst, saying of an appeal:

> If it fails it means that all I have done is to prove that though in Protectorates the inhabitants owe the Government an obedience which can be enforced, as foreign subjects they have no access to the courts. Though 'protected' they have less protection than British subjects since no act of the executive is subject to review. Constitutionally that is an impossible situation but being both *de facto* and *de jure* it will be long before strong enough forces are marshalled to change it. I can do nothing more now ... .[15]

Leys supplied information to an anonymous 'learned friend' who wrote an article for the *Glasgow Herald* on the implications of the Maasai decision for the legal status of protectorates. Sending Harvey the cutting, Leys said he was 'arranging for the subject to be treated more extensively in the legal press'. As things stood, 'the Crown has no conceivable limits to its sovereignty but the subject *de facto* has no shred of rights ... The law lays upon [the Maasai] the duty of obedience but denies them the name and the privilege of subjects.'[16] The *Herald* article focused on the nature of sovereignty, pointing out that 'a Colonial protectorate has been described as a form of control, falling short of full sovereignty, assumed by a civilised State over the territory of an uncivilised or semi-civilised community'. But, the writer contended, Britain was in fact exercising full sovereignty in East Africa, manifested in its Crown appointees, legislation and penal code, and raising of revenue through taxation. It also ran the Uganda Railway as a state undertaking and maintained troops and police. As the law stood, the term protectorate connoted foreign territory, to which the powers exercised by Britain in East Africa were inapplicable. Former protectorates, such as Bechuanaland, could become dominions without formal annexation, by virtue of the fact that large numbers of British subjects had permanently settled there with the assent of the Crown (a situation referred to in *Rex* v. *Earl of Crewe*). Leys believed the plaintiffs would pursue this line of argument before the Privy Council, but the case never got that far.

At the Colonial Office, Permanent Under-Secretary Sir John Anderson privately said of the judgement: 'I do not like the decision at all ... to call the Agreement a Treaty is an abuse of language.' He could not imagine the Privy Council supporting the judgement if the case was referred to it.[17] A legal advisor, C. Tennyson, thought it 'an ingenious way of evading a decision on its merits'.[18] Earlier, Tennyson had expressed pessimism about winning the case at all, while Harcourt had been advised by a predecessor, Alfred Lyttelton, now a Privy Councillor, that if it went against them they should settle with the Maasai in cash. In his memoirs, Frederick Jackson darkly noted: 'There is one thing about the Masai move that is not known, or only by very few, and that is that Mr. Alfred Lyttelton was fully cognizant of all the facts and details of it, when he arrived in Uganda and stayed with us. He told me that he might be placed in an unpleasant position on his return home, as he would have to say what he knew, as a Privy Councillor, and not as an ex-Colonial Secretary.'[19] What he knew was not spelled out. Significantly, the Colonial Office clearly anticipated as early as January 1913 that the government might lose the case, and began laying plans to take it to the Privy Council if necessary.[20]

The case went to appeal in December that year before Morris Carter (who went on to chair the Kenya Land Commission in the 1930s), Bonham Carter and King Farlow in the Court of Appeal for Eastern Africa.[21] The earlier decision was upheld and the action dismissed. The Crown restated its claim that the two Agreements did not constitute legal contracts between the Protectorate and the Maasai signatories. The second treaty was termed no more than a 'modification' of the first. Morrison, for the Maasai, argued that the position differed from what it was 30 years before. British rule and courts had now been established, and the Maasai were not foreigners in the courts but equal to British subjects in every way:

> A treaty can only be entered into with an independent Sovereign State, the chief of which is not subject to the jurisdiction of the Courts as is the Chief of the Masai ... If the Masai took up arms against the Government they would be rebels, liable to penalties for treason, that is to say they have the liabilities and equally the privileges of subjects.[22]

There was in fact, said Morrison, no difference between the Protectorate and a Crown colony. But a protected foreigner was still entitled to full protection of his rights in English law. The Agreements were civil contracts, enforceable in the courts, and not unenforceable treaties. And to take the Crown claim about Acts of State to its logical conclusion, he argued that a squatter refusing to leave the land reserved for the Maasai could only be ejected by an Act of State. Attorney-General R. M. Combe, appearing for the Crown, Hemsted and McClellan, admitted that if the Maasai were British subjects the plea of Act of State would not stand up. He argued that the Crown recognised 'a concurrent jurisdiction with that of our courts in some of the native chiefs'. This indicated that there was a 'material difference between the position of a British subject and of a native of the Protectorate'. Combe dismissed the appellants' claim that they were not subject to Crown-appointed chiefs and that the signatories had no business making a treaty with the Crown: 'A native member of a tribe is a subject of the head of the tribe, whether the "head" be an individual or more persons than one, and it is for the Crown alone to say with whom it will make a treaty.'

Morrison made other points, including: Maasai sovereignty had never been recognised and did not exist now; the exercise of law by chiefs was only by consent of the people and did not imply sovereignty; the Crown had assumed ownership of minerals and granted rights in land, and annexation would merely entail a formal recognition of the existing state of affairs; the 1910 Petition of Rights Ordinance gave the Maasai as much right to sue the government in the Protectorate's courts as it gave British subjects the right to do in Britain; the Crown was claiming in this case to do by prerogative powers what it had power to do by legislative act; and full British constitutional rule had been established under the East Africa Order in Council 1902.

In his part of the judgement, Morris Carter maintained that a protecting state could permit some vestige of sovereignty to remain in the 'native' authority, and it must be taken to have permitted this, unless it had assumed full sovereignty by formal annexation. The latter had not happened, and no one could force the Crown to take territory it did not want to acquire. In his view, BEA was still a foreign country. The granting of a system of laws did not necessarily imply that a constitution had been introduced, and the Order in Council 1906 had not provided East Africa with one. In asking whether the Maasai were a people with whom a treaty could be made, he turned to considering their status. He concluded that they were not British subjects but 'subjects of their chiefs or their local government, whatever form that government may in fact take'. They were still a tribal entity with whom the government could make a treaty if it so wished, and there were other precedents in East Africa (the Ankole Agreement of 1901 in Uganda) for the making of a treaty with a 'tribe ... under the active administration of the government of a Protectorate'. There was 'a remnant of sovereignty still remaining in the Masai', as evidenced by official recognition of the power of tribal chiefs to try legal cases. He rejected the suggestion that the treaties were merely contracts, and said the court could not go into the question of whether the government had selected the right persons with whom to make treaties. He believed it was also beyond the remit of the court to investigate whether or not a treaty had been obtained by duress. Anyhow, he claimed, duress was a common feature of treaty making, so that was nothing unusual. Furthermore, the Maasai had no rights to claim against tortious acts committed in the name of the Crown. Quoting Vaughan-Williams in *Rex* v. *Earl of Crewe*, Morris Carter said:

> The idea that there may be an established system of law to which a man owes obedience, and that at any moment he may be deprived of the protection of that law, is an idea not easily accepted by English lawyers. It is made less difficult if one remembers that the Protectorate is over a country in which a few dominant civilised men have to control a great multitude of the semi-barbarous.[23]

The Crown had laid bare its race and class supremacism, justification for any action it chose to take. Morrison might have queried the self-compellant nature of the phrase 'have to control'. He could also have pursued the question of whether the protection of 'native' inhabitants was not the first concern of a protectorate, if not its *raison d'être*, as the Colonial Office had repeatedly implied. King Farlow seemed to contradict this:

> It was obvious that the Masai, with their roving habits and warlike traditions, were not desirable neighbours for white settlers, and that their presence along the recently constructed railway was hardly consistent with the public interest ... Lord Justice Farwell, in ... his judgement in *Rex* v. *Earl of Crewe*, points out that when the State takes over the responsibility of a Protectorate over a territory inhabited by native tribes who largely out-number the white population, its first duty is to secure the safety of the latter. I am of the opinion that the Government in discharge of this duty was compelled to enter into this agreement as an act of high policy or necessity.[24]

Who was being protected from whom? As Ghai and McAuslan later commented: 'A system of rules which weighed the balance so heavily against those most in need of its protection must be regarded as a very dubious benefit indeed.' Moreover, it appeared from this case that 'a British protected person is protected against everyone except the British'.[25]

Most importantly, the Crown contradicted itself yet again over the question of who wielded political power in Maasai society. Morrison failed to pick up on this. On the one hand, the Crown stated that the treaties arose out of conferences between 'certain of their chiefs' and the government, and were entered into by 'the Chief, (his regents) and certain representatives of the Masai tribe'. On the other, it acknowledged in two separate passages that power lay with the warrior class, while prophets only had spiritual authority:

> owing to the nomad and truculent nature of the race the central authority had not a very great power; the main power rested with the warriors, who gave utterance to their wishes through their elected chiefs. There was also an individual called Laibon ... [who] ruled more as a spiritual than a temporal chief.
>
> ... The ruling authority among the Masai is represented by their warrior class, with elected chiefs and a chief medicine man or Lybon acting as an advisory and semi-controlling force.[26]

As for the status of BEA and hence the position of the Maasai within it, Morrison could also have taken issue with Morris Carter's assertion: 'It has not been argued before us that East Africa has been acquired by settlement, nor has the Court been asked to take any evidence on this point.' Since settlers were clearly not temporary visitors, evidence of permanent settlement might have gone some way to proving that BEA was *de facto* annexed territory and the Maasai were British subjects. This point was later made by Buell (see below). Curiously, the Crown ruled that the treaties were a moral rather than legal matter: 'The obligations imposed here on both parties were moral, not legal' – an interpretation reiterated by the British government in the early 1960s at pre-independence talks.

The Maasai were given conditional leave to appeal to the Privy Council, but this lapsed when they failed to give security for costs. But, if one is to believe the stories, there was another, hidden reason for Maasai failure to appeal to this higher authority – Ole Gilisho gave up after being threatened with drowning at sea if he set sail for Britain. This is how Ole Mootian described it to me:

> Ole Gilisho was told that his appeal was overseas and on a certain date. So he went with a train to Mombasa. There he was told by an old man called Marianyie Ole Kirtela: 'If you go, when you reach the middle of the sea they will drop you into the water. Then your case will be finished. So you, Maasai, don't go!' It was a white man who was telling the other one [Ole Kirtela] to tell Old Gilisho that.

Ole Kirtela was allegedly paid to pass this message on. If true, this is devastating news, and explains a great deal about Maasai failure to pursue the case. It also fits with the pattern of intimidation of Ole Gilisho by government officials. The story was repeated by two other informants, Joseph Ole Karia and Nkaburra Ole Njapit. But as with many stories of this nature, it could also stand for social metaphor (Ole Gilisho's sense of insecurity outside his own territory) and wishful thinking by later generations of Maasai (the London-bound delegation might have succeeded if only it had not been hampered at the outset). Also, it is not consistent with the fact that Ole Gilisho had become a defendant in the case; why should he, not Ole Nchoko, have considered appealing to the Privy Council?

Morrison wrote a postscript to Harvey in June 1914, calling into question his clients' reliability:

> I regret to say that the Masai appeal home has come to an end owing to the Masai failing to find security for the costs of the appeal. On the

whole I am inclined to believe that this is the result of the long delay and constant statements of the Government officials that they had no chance of success. Some of them tell me that they are very dissatisfied and would like to go to the expense of a special order of the Privy Council giving leave to appeal out of time but I doubt very much if they have the necessary energy to arrange for this … I doubt very much if it is worth while pressing for a commission at this stage because of all the unreliable native evidence, Masai evidence is the most unreliable. Of course had the case gone on to be heard on the merits the technical points could have been established by Hollis's book among others, and then it would have been for the Government to prove that individual Masai had agreed to the move.[27]

Morrison later sued the Maasai plaintiffs for unpaid costs, but lost in the High Court. Ole Gilisho reportedly offered his ring as surety, and Morrison refused to give it back until the fees were paid.[28]

Critiques of the case

Dissident game warden George Goldfinch summed up informed 'liberal' opinion in a letter to the Anti-Slavery Society in the 1920s: 'I think it was at the time rather generally considered that if the case had ever been allowed to be heard that the Masai would have won hands down because that [1904] agreement had been drafted by the very people who would have heard the case and was intended to be unbreakable.'[29] In his book *Kenya*, Leys commented:

It is scarcely necessary to remark that Masai sovereignty is a legal fiction to which there is nothing whatever in fact to correspond. Masai taxes are paid to, Masai murderers hanged by, the British Government. The legal status of a protected tribe, in fact, gave to the Government the power to do exactly as it pleased, power greater than any Government over British subjects can acquire, while it debarred the tribe from acquiring any rights in land of which a British Court could take cognisance.[30]

Twenty years later, the Kenya Land Commission (KLC) heard evidence about the case in a memo submitted by P. D. Master. He cited Professor Raymond Buell's 'instructive' criticism of the judgement, which is worth quoting in full:

Had the British authorities made a contract in 1904 with a European settler granting him certain land, the contract would have been enforceable in the British courts. But according to this decision, an agreement made between the British authorities and the representatives of some forty thousand natives was not enforceable by the courts. If the Masai nation really had an international status as a state, no objection to this decision might legally be taken. But in the case of East Africa, the British had extended a judicial system throughout the country and it had erected a Legislative Council, the acts of which the Masai were obliged to obey. Their consent to these acts was as tacit and as fictitious as the consent which Rousseau's happy savage gives upon entering the social compact. A few years later, the Privy Council decided that despite the fact that Southern Rhodesia had never been actually annexed, it was in effect annexed to the Crown, because of the permanent occupation which has been established throughout the country. If the same argument had been followed in the Masai case, the court would have held that in view of the permanent European settlement in East Africa, it had become in effect part of the Crown's dominion, and that the Masai were therefore entitled to the guarantees of British subjects. If the rights of the Crown were as limited as this judgement implied, it would appear that both the Crown Lands Ordinance of 1902 and that of 1915 authorizing the alienation of land were *ultra vires*. In the protectorate of Sierra Leone, it had even been held that, for the purpose of taking an oath of allegiance to the Legislative Council and of joining the army, a resident of the protectorate is a British subject. Thus the Masai judgement appears to be inconsistent with the opinion of the Privy Council in the Rhodesian case. It is a curious fact that both decisions conform to the interests of the European instead of the native population.[31]

The memorandum went on to cite a 1921 judgement, *Gathomo* v. *Indangara*, which held that 'natives' were tenants at will of the Crown since native areas were Crown lands.[32] This decision implied that trustee governments had the right to confiscate the lands of their wards, without compensation or the right of appeal: 'Such a right, masquerading under the forms of law, is worse than the right of the robber-barons, for it gives the right (and the right has been used by the Kenya Government) to the guardian to rob the ward.' Since Britain had declared in a 1923 White Paper that the principle of trusteeship for the natives was 'unassailable' in Kenya, and the Permanent Mandates Committee had expressed the opinion that mandatory powers did not possess any right over any part of a mandated territory other than that which resulted from being entrusted with its administration, it had a straight choice:

> either [Britain] can elect to enjoy the rights of free plunder given it by decision of its High Court in Kenya, and from time to time run up the 'Jolly Roger' in place of the Union Jack, or stand by its solemn professions and deny both now and retrospectively that it has or had any rights over any part of the territory of Kenya, other than those arising from its assumption of the administration of the territory.[33]

The memorandum concluded that land policy in Kenya 'has been and is fundamentally wrong and suicidal, both for the Government and the governed. Moreover, a proper inquiry into the land policy pursued in Kenya would reveal that native lands have been appropriated by the Government, ignoring all considerations of the natives.' It referred readers who might want to know more to the works of Leys and McGregor Ross. Morally and legally, it considered that such lands still belonged to Africans. If justice was to be done, all lands wrongfully appropriated by government and given to settlers should be returned to Africans, with full compensation including funds for their development.

McGregor Ross also gave evidence to the KLC, both in person and in a memorandum. He cited a legal treatise which stated that the establishment of a protectorate did not entitle the protecting power to deal with private rights to land in the territory; any such power must be granted by the local government. Second, 'the native inhabitants of a Protectorate have not become the nationals of the protecting State, and that State cannot validly compel them in any particular way or deal with their property, unless the right to do so flows from one of the transferred powers.'[34] Ross said no powers had been transferred to Britain authorising it to deal with African-owned land without African consent. He had discussed the issue several times with Jackson, who said none of the treaties made by the IBEAC with African tribes 'contained anything that was intended to mark, or could be construed as marking, any transfer of control over land in African ownership or occupation out of African hands'. On the contrary, the Company's charter required it to show 'careful regard' to African lands and property. Ross also quoted Lord Lugard as saying that, in countries acquired by conquest or cession, 'it has been laid down as a principle from which no civilized Government would think of departing ... private property, whether of individuals or communities, existing at the time of cession or conquest is respected.'[35]

Ross believed the government of the day had been well aware that the validity of its actions over African land was 'to say the least, dubious'. In *Kenya from Within* he had referred to a proposal by Lord Crewe, when Secretary of State for the Colonies, that the government should make treaties with every tribe in the Protectorate, acknowledging their 'absolute possession' of land in tribal use or possession. Jackson had written to Ross about this in April 1925, lamenting that no action had been taken; he urged Ross to 'whip up all your MP friends to take up the matter ... and urge the present Government to adopt the suggestion'. Crewe also proposed that all native reserves should be demarcated by government officials and representative Africans, and that once treaties or agreements had been approved by the home government, all steps should be taken 'to render it impossible for the local Government to in any way alter a clause without the consent of Parliament'. Jackson could not remember the date of the despatch, but urged Ross to get hold of it. Ross hoped the KLC would act upon this and, quoting Jackson, restore 'our reputation for fair dealing'. It did nothing of the sort.

Thirty years later, Claire Palley discussed the broader issues around the status and development of protectorates, in a useful chapter that mentioned the Maasai Case.[36] Ghai and McAuslan concluded that the judgement amounted to saying that the Maasai retained sufficient sovereignty to make a treaty but not to make a civil contract about land, and that the case 'points to a paradox of power in a protectorate'. In such territories, 'there is a residue of sovereignty left to the protected people or state and the Crown does not have unlimited jurisdiction therein which is the case in a colony where the Crown has sovereignty, yet the general law relating to the actual operation of this jurisdiction has the effect of conferring a greater immunity on the colonial administration in a protectorate than in a colony.'[37] More recently, David Williams, in a study of New Zealand's Treaty of Waitangi, concluded that 'the judgements in this [Maasai] case are noteworthy for the adeptness of judges in arriving at appropriate legal doctrines to legitimate the spoliations of a colonial Government ... The reasoning of the judges was remarkable for ingenuity in ensuring that the plaintiffs obtained no remedy.'[38]

What about Maasai opinion? Unfortunately, surviving elders' knowledge of the 1913 case is scanty; the generation that knew the most has passed away. But on the available oral evidence the case does not appear to have been widely discussed at the time or subsequently, except in leadership and legal circles. Said one informant: 'It was not talked so much about by the community.' When asked what happened in court retired councillor Ole Yiaile, whose father came south with Ole Gilisho, replied: 'Nothing, but we guessed that they were defeated because when they came back they did not mention anything to us.' Others said: 'It just faded away.' The general perception was that this was a white men's court, therefore it was no surprise that the decision went against the Maasai. Modern day legal opinion will be given in the Conclusion.

Once the 'northern' Maasai had been moved to western Narok, grievances expressed in the plaint and by sympathisers continued to resonate in the early decades of the twentieth century. In particular, allegations that the community and its livestock were being exposed to diseases to which they had little or no resistance, and fears of an impending environmental catastrophe, appeared to become a reality – at least from a Maasai point of view. The following chapter will explore the wider environmental impacts from different perspectives. Sandford merely commented: 'The Masai did not, as a whole, settle down contentedly in their new surroundings and they showed a disposition to make the worst of everything. Complaints were incessant that the country was unsuitable for them and for their stock, and every device short of active resistance was employed to obstruct the Administration and to defeat the ends of justice ... The "Masai Case" was undoubtedly responsible for much of this feeling ...'[39]

5
The Ecological Impacts

Shomo pii Ngatet, enakop o ol-tikana
Go completely to the south, land of East Coast fever and malaria.
What Maasai believe the British told them,
while forcing them from Laikipia

More than 90 years after the second move, Maasai elders in western Narok still talk with passion about its effects on the health of humans and herds. They describe the impact of the move in 'pathological' terms, believing that the British deliberately sent them 'to that land where *ol-tikana* is' in order that they might die there.[1] They claim that they and their herds succumbed to diseases in the Southern Reserve which were unknown or not prevalent in their northern territory, most specifically Laikipia, and that they have been blighted by sickness ever since. They insist that the land they were moved to was not only grossly inferior to Entorror in terms of water, grazing, ticks and tsetse fly, but that the new environment infected and killed them. It was literally deadly. Some go further, and insist there was *no* disease in Entorror. In the collective oral mythology, Entorror is seen as Eden, its sweetness constantly compared to the bitterness of the south, or Ngatet.

Their quantitative land losses in this period are well known; it is generally acknowledged that the Maasai lost more land to the British than any other ethnic group in East Africa. But their qualitative losses, in terms of the comparative richness of their northern habitat and their alleged propensity to disease in their new environment, have not been examined in detail. Sindiga has described how 'colonial intervention in Maasailand led to the breakdown of traditional ecosystems', and attributes the subsequent severe degradation and population pressures of Kajiado and Narok Districts to a process begun in 1904.[2] Tignor writes: 'The loss of the dry weather grazing forced the Maasai to overwork the more arid southern reserve, resulting in loss of vegetation, soil erosion, and overall decline in grazing. In the twentieth century the reserve became progressively less able to support its livestock.'[3] Rutten, in a study of factors that have led to land losses and undermined the livestock economy in Kajiado District, comments: 'Moreover, in terms of quality the loss [of alienated land before 1912] was even more severe as green pastures located in ecologically favourable areas had to be abandoned and were replaced with a less comfortable habitat, heavily infested by tsetse fly and mostly lacking sufficient water and all year round grazing.' Western and Nightingale have made the connection between the early land alienation and Kajiado pastoralists' increased vulnerability to drought. Other scholars have made similar remarks, briefly linking the early losses to later degradation, drought and other pressures.[4] This chapter will attempt to explore the qualitative 'before' and 'after' of the land alienation in more depth.

In theory, the area of land to which the Maasai were moved, the 4.46 million-acre western extension added to the Southern Reserve in 1911, seems generous until one examines its quality. (The reserve as a whole was nearly 10 million acres, and various additions and excisions were later made.) Also, it is not simply a matter of the quality of the land and other natural resources. Such is their dependency on livestock, and their total identification with cattle in particular – *en-kishu* means both cattle and the Maasai as a people – that cattle disease in any environment is inextricably linked to human health; stock and human health are spoken of almost interchangeably. This fact will be implicit throughout this chapter. Furthermore, the grievances of this group of migrants must be seen in the context of acclimatisation over time and space. There were of course other Maasai already living in this area when the 'northerners' arrived – members of the Loitai, Damat, Keekonyokie and Siria sections, as well as a quarter of the total Purko population – and so it cannot be described as an environment in which Maasai could not survive. The point is that the newcomers were unfamiliar with this environment, took time to discover and experiment with a different range of wild foods and medicinal plants to those they were used to in Entorror, were non-resistant to certain infections, and in the interregnum between arrival and acclimatisation, when some resistance would have developed, both humans and stock suffered acutely. It is this suffering which colours people's memories of the move and what happened immediately afterwards.

There were contemporary parallels in German East Africa, where amateur ethnographer and colonial official Moritz Merker observed the painful 'acclimatisation process' which the Maasai underwent after the epidemics and civil wars of the late nineteenth century.[5] Many were forced to move from a nomadic to a sedentary lifestyle, from pastoralism to agriculture with its accompanying change of diet, from one climatic zone to another. This resulted in 'a great sacrifice of human lives', greater susceptibility to disease and a reduction in women's fertility. Merker did not make an explicit link between forced moves, the resulting loss of medicinal and nutritional plants (the therapeutic uses of which he details in a fascinating appendix), and sickness. I suggest that this was a vital part of the equation in forcibly displaced communities on both sides of the border.

Some analytical challenges

One could attempt to establish whether there is any scientific or biomedical basis for Maasai claims that the Northern Reserve was effectively ECF-free, and of a deliberate 'move to kill' policy driven by administrators' knowledge of the presence or absence of disease, particularly ECF, in the two environments. There is some compelling evidence to support the first of these claims, while some colonial officials certainly 'subscribed to a Malthusian view of disease and drought as the natural regulators of the [African] stock population ...'.[6] But the search for scientific evidence is also an unsatisfactory exercise, in part because early scientific data simply does not exist, and because the exercise involves comparing like with unlike: to put it crudely, a Western scientific view of disease which is rooted in diagnostics and laboratory experiment, versus a more holistic indigenous view which regards 'dis-ease' as a natural part of life.[7] Both systems of thought and practice have their own taxonomy, within which there are some points of agreement – for instance, Maasai used inoculation long before European contact.[8] But diagnosis and treatment are often so completely different that it is difficult, at times, to know whether one is dealing with the same disease. The dangers of sharply demarcating an ecologically harmonious 'before' and disharmonious 'after' colonial intervention in Africa are also very real.

Most importantly, the subject is larger than scientific: it concerns disease as a metaphor for colonial encounters, and what these produced in social and other terms. In this case, I argue that ECF has come to represent – for the older generation of Purko Maasai at least – infection by colonialism, and it is their conceptualisation that interests me. Therefore I aim to examine what scientific and quasi-scientific evidence there is in tandem with *perceptions* of disease, and perceptions of diseased versus healthy environments, confining my focus to the Purko of western Narok, with other sections mentioned briefly at the end. Although other diseases affected Maasai stock in both reserves, I shall focus largely on ECF, partly because this was repeatedly raised in oral testimony. This is strange, given the high incidence of fatal BPP and other diseases in this period. Several reasons were suggested for this: ECF killed stock faster than, say, trypanosomiasis, there was less 'natural' resistance, and it was particularly fatal to young stock, the bedrock of the future herd.[9] It may also be that ECF is recalled as an extreme threat because, compared with other stock diseases for which vaccines were developed early on (they were available for rinderpest and BPP by 1912), there was then no vaccine for ECF; it was only developed relatively recently. There was nothing that the colonial vets could do for ECF, besides advocate dipping with acaricides to kill the ticks, or hand-dressing, which involves applying an oily dressing to the poll, ears and under the tail. In the early days, temperature camps (a model developed in South Africa) were used to isolate infected herds; indigenous treatments will be mentioned later. As for the human variant of *ol-tikana*, 'malaria' may be a catch-all term that encompasses fevers and sleeping sickness (called *en-kasilei*, also the Maa word for tsetse fly). In the early stages sleeping sickness can be mistaken for malaria or flu, even by doctors.

Both parties to this encounter saw the other as a pollutant, who had brought or was believed likely to bring diseases to humans and cattle. Blaming foreigners for introducing disease is, of course, a common theme in the history of syphilis, HIV/AIDS, plague and other epidemics. These particular perceptions were embedded in nineteenth-century encounters between travellers, traders and Maasai. By the end of the century, the plagues that devastated Maasai society reinforced this view of ill-omened and infectious newcomers. Colonial administrators' attempts to curb cattle numbers, and the judgmental rhetoric around pastoralists' alleged 'cattle complex' and 'bad' management of stock and pastures which continues to this day, soon multiplied Maasai suspicions of British motives. The feeling on Laikipia well before the second move, according to DC Kenneth Dundas, was 'that Government is mainly anxious for their cattle to die'.[10]

'Clean' and 'dirty'

Veterinary officials demarcated BEA into 'clean' and 'dirty' areas, according to the incidence of ECF. This is how they described it in 1911–12: 'The whole country may be divided into two areas which must be classed clean and dirty. The clean area, somewhat in the form of a wedge or "V", cuts into the infected area, nearly dividing it. The broad base of this area extends northwards towards Abyssinia ... Passing southwards, Laikipia is reached. This area, the grazing ground of the Masai, contains a large number of cattle which are free from East Coast fever.'[11] It was deemed to be clean. The Rift Valley, where most of the imported settler stock could be found, was also at that stage free of ECF. The dirty areas stretched west into Uganda, south to the German border via the Southern Maasai Reserve, which was classed as dirty although only 'slightly infected', and east all the way to the coast. In these areas, ECF was said to exist in varying degrees of endemicity. In the official lexicon, 'clean' and 'dirty' generally referred to areas of healthy or sick cattle populations and areas of white or African settlement respectively, and underpinned administrative action to keep them separate. This ignored the fact that some settlers' farms in supposedly clean areas such as the Rift were actually dirty, as Norval, Perry and Young point out:

> As time went on and people's understanding of the disease increased, it became apparent that the disease was more widespread than originally thought, and the simple categorisation of the country into clean and dirty areas on the basis of the presence or absence of 'native' stock was not always valid ... outbreaks of ECF in the clean area [of the Rift] were not uncommon.[12]

Social control of Africans ran parallel with these demarcations and implicitly justified their introduction. The regulations also favoured settler production; although quarantine hampered white farmers too, it effectively blocked African producers' attempts to enter the market economy.

The administration tended to view 'native' cattle as inherently inferior and diseased, though veterinarians recognised their resistance to some diseases, including ECF.[13] It therefore sought – through compulsory measures such as fencing, quarantine and removal of squatters – to keep 'native' stock separate from imported, pure-bred settler stock. But some leading settlers, notably Gilbert Colvile, took a different view of Maasai cattle (a type of small East African Zebu) and used them as the basis for his breeding stock. As his former farm manager, Desmond Bristow, told me with pride: 'He bred a type of animal which originated from the Maasai, not from the Northern Frontier Borans. They were improved Maasai and very good cattle too.' Bristow still has remnants of this original herd, 'rescued' when Colvile's 'Ntapipi' farm was sold.[14]

For their part, the 'northern' Maasai also saw the Southern Reserve as 'dirty'. They believed, both before and after the moves, that they were deliberately moved to inferior and waterless pasture infested with ticks and tsetse fly, where both human and stock disease were rife. The British had promised the very opposite, when they urged relocation to the south on the grounds that the Maasai would be 'safer' there, defining safety as freedom from cattle disease. Significantly, as mentioned in the previous chapter, much of the 1913 plaint was devoted to concerns about human and stock disease in the reserve and *en route* to it. It specifically cited ECF: 'The Southern Masai Reserve to which the stock of the Masai is being moved is infected with East Coast fever ....'

Comparing the two habitats

Maasai elders describe their movement from the north as 'dropping down', and typically say 'we came dropping', which graphically expresses the geographical facts. Taking Entorror to mean the whole of their northern grazing grounds, but most recently Laikipia, the Maasai swapped a territory that was generally higher, cooler and wetter for largely semi-arid plains with few highland drought refuges apart from the Mau, the Loita Hills and other highland areas in and close to Trans-Mara.[15] Sindiga notes: 'Only a small fringe of territory had good grass and water all year round. Nearly all permanent streams were controlled by European settlers. The rest of the territory within the reserve was either without water, or contaminated by disease, or in European control. At their level of technology, the Maasai could not readily use a total of 51 per cent of their reserve.'[16]

Although Laikipia was not ideal and not large enough, it offered a wider variety of options than the western extension of the Southern Reserve, particularly at times when policing was relatively relaxed and herders were allowed to cross boundaries during droughts. Leys compared the two territories in this way: 'No European in the country imagined for a moment that the Masai on Laikipia wished to leave it. The area, though small, is as fine a piece of country as there is in Kenya, with rich soil and perennial streams, vastly superior in every way to the country south of the Rift Valley ....'[17] The Southern Reserve included large areas that were 'arid and useless'. Leys identified a major drawback: 'The great defect of the new reserve is that all the streams in its western half rise in a mountain mass [Mau] that then belonged and still belongs to two rich Europeans. Only a small fringe lying just below the mountain had perennial water and good grazing all year round.' The two farmers were Powys Cobb and Lord Delamere. Some of the 'streams' were, however, more substantial than his term suggests; they included the Siyabei, Uaso Narok and Uaso Nyiro rivers.

In leaving the highlands, the 'northern' Maasai lost the wide choice of habitat they had enjoyed up until 1911; until 1904, of course, the choice was even wider. Transhumant pastoralists make use of a great variety of ecological niches. If they are free to do so, and control the territory, they move in and out of these niches according to seasonal need, and constantly stress the balance to be achieved in rangeland and stock management between highland, dry-season grazing (*osupuko*, the drought refuges) and lowland, wet-season grazing (*ol-purkel*). Moreover, each section has its own *osupuko* and *ol-purkel*, so they cannot easily find alternatives on moving to a region already occupied by other sections. Highland drought refuges are vital not only for dry-season grazing, water sources and salt licks, but also because they offer pastoralists strategic control over surrounding wet-season pastures and major stock routes.[18] Many include forested areas, which contain important medicinal plants; the Maasai word for tree (*ol-chani* or *ol-cani*, pl. *il-keek*) also means medicine. Wetlands in relative drylands, such as the swamps east and north of Lake Ol Bolossat and those bordering the Uaso Narok river north-east of Rumuruti, were also very important, biodiverse dry-season resources. Such wetlands have been described as 'the most productive ecosystems in the world'.[19]

In western Narok there were no swamps to speak of, and those in Chepalungu were out of bounds. Compared to the north there were few accessible forests, apart from those on the Mau, those south-west of it and at Chepalungu, but these were a long way from the best plains grazing. Also, Mau and Chepalungu Forests were truncated by the reserve boundary; the best part of Chepalungu was not in the reserve at all. At the time, no proper forest surveys had been made, although there had been a Forest Department since 1902 and a Forest Ordinance became law in 1911. However, D. E. Hutchins (later Conservator of Forests) wrote reports in 1907 and 1909 based on his observations of woodlands, including those made in a two-month trip to the Mount Kenya area with McGregor Ross. His 1909 map did not mark any forests in the Northern Reserve, apart from the northern Aberdares on their southern fringe, and neither did it show the extensive Chepalungu in the south, so it can hardly be called accurate. In 1920 a more detailed report on forests and timber resources was published, followed two years later by that of R. S. Troup of the Indian Forest Service, based on inspections made in 1921. A map in Troup's report clearly demonstrates the difference between the two environments.[20]

A clearer picture emerges in the 1960s, with a *Catalogue of the Forests of Kenya*.[21] For Laikipia and its environs, this lists the forests of Ol Bolossat, South Laikipia, Lariak, Marmanent (*sic*, Marmanet, north of Thomson's Falls), Mukogodo, Ndare, Ol Arabel, Rumuruti and Uaso Narok. The total is 212,245 acres of available forest in Maasai territory or very close to it. By comparison, the forests listed for western Narok and Trans-Mara are Chepalungu (25,174 acres, of which maybe a quarter lay inside the reserve), Ol Posimoru Forest north of Narok town (91,300), and Trans-Mara Forest (84,968), totalling 182,561 acres.[22] However, large areas of this woodland were inaccessible and a relatively long way from where the incoming Maasai gravitated to pasture and water.

In the absence of early scientific studies one has to turn to other government-sponsored reports on fauna and flora, and travellers' accounts, in order to build up a picture of the early environment of Maasailand, however unsatisfactory and incomplete this may be. Thomson described the stark differences between the two areas of what would become Kenyan Maasailand. He did not travel as far west as today's western Narok, and the southern deserts depicted here (the Nyiri plain, and the dry bushy area of 'Dogilani' or Iloodokilani, named after the Maasai section) were not typical of the greater part of the country to which the Laikipia migrants came. Even so, references to rainfall and dry-season grassland are broadly applicable.

> The Masai country is very markedly divided into two quite distinct regions, the southerly, or lower desert area, and the northerly or plateau region. The southerly is comparatively low in altitude ... from 3000 to nearly 4000 feet. It is sterile and unproductive in the extreme. This is owing, not to barren soil, but to the scantiness of the rainfall, which for about three months in the year barely gives sufficient sustenance to scattered tufts of grass.[23]

By contrast, the northern region was totally 'charming', and featured a 'very network of babbling brooks and streams'. Rainfall on the Laikipia plateau was estimated at 30–40 inches per annum compared to 15 inches in the southern region.[24] Coastal Africans in his party could not stand the damp cold and air at these higher elevations, but Thomson found them invigorating. He was well aware of pastoral seasonal migration, and the necessity of 'moving up from the plains to the highlands in the dry season and vice versa in the wet season'.[25]

Harry Johnston, describing the eastern province of the Uganda Protectorate in 1902, just before it was transferred to BEA, wrote of its southern portion: 'south of Lake Naivasha the territory ... stretches in a narrowing angle towards German East Africa, and the country in this direction becomes increasingly arid and lacking in rainfall.' By comparison, the delightful northern Rift which he recommended as 'the future white man's colony' was 'a country of rolling grass-lands, dense forests of confers, and bamboo-covered mountains'.[26] Laikipia itself was just outside the province boundary. Later descriptions of the natural resources of the Rift Valley, the Mau and Laikipia in the early 1900s included those by Eliot, the Agricultural Department and Hobley. Eliot described in glowing terms 'the beautiful little plain of Endabibi' northwest of Naivasha which Colvile was to make his prize farm; the good water sources, grazing and rich loam on the southern Mau; the Morendat, Gilgil, Magalia and Enderit rivers flowing into Lakes Naivasha and Nakuru; the excellent pasture around Enderit and 'almost everywhere in the Rift Valley'.[27] This included the much-prized stargrass (*Cynodon dactylon*).

The area around Njoro, site of Lord Delamere's first farm, was said to be the finest part of East Africa. Why was the grazing so good? Because the Rift 'has been continuously grazed by native cattle during many years'. Agricultural reports in 1904 also noted that the 'quality of the pasture in the Rift is entirely due to grazing', and praised the 'great drought resisting power' of its grass species.[28] Hobley, in discussing how the Maasai should be compensated for losing their grazing rights in the Rift, regarded its riparian areas as particularly desirable and recommended that 'the rich evergreen meadows in immediate proximity to the lakes and which are a great standby in periods of drought [ought] to be assessed at the highest rate'.[29] When Hobley toured Laikipia in June 1904 to check its suitability as a reserve, he reported from the top of the escarpment:

Looking East the country is a boundless green rolling plain ... with belts of thick forest in every valley ... generally speaking the plains appear to be devoid of game. Next day we marched South-South-East through magnificent grazing country [and] passed three sources of water, swampy streams [called by the Maasai] Ol-are loo-naitolia, Ol-are loo-'l-Torobo, Ol-are loo-'l-Morijo [which] all appear to drain towards the Euaso Narok ... a fine stream twenty to thirty yards wide.[30]

Everywhere he turned there was thick, luxuriant grass; waterholes, waterfalls, streams, rivers and a lake; 'beautiful belts' of forest; and little game to trouble people outside the forests in daytime, although buffalo emerged at night to graze. The land only became 'somewhat drier' as he marched towards the Pesi Swamp. Dropping down to 'the famous grazing grounds of Ongata Bus' he could see Lake Ol Bolossat – 'a big fresh water lagoon' from whose northern end the Uaso Narok river flowed. This actually consisted of two lakes about a mile apart with a connecting swamp. The report confirms what the Maasai claim today – that there was plenty of water, wonderful pasture, little or no big game compared to the southern plains and therefore fewer tick hosts.

In the south, by comparison, the incomers began to face stiff competition from wild animals for pasture and water. There were larger numbers of predators and disease-carriers. For example, wildebeest migrating north from the Serengeti were a walking reservoir host for malignant catarrhal fever, transmitted to cattle in the three or four months following wildebeest calving; even by the 1960s the death rate was more than 95 per cent and no treatment or vaccine was available.[31] The only permanent rivers were the Mara, its tributary the Talek, and the Uaso Nyiro; other watercourses dried up in drought years. Today, 'rainfall is erratic both in amount and timing'[32] and rivers such as the Uaso Narok, Lemek and Siyabei dry up in drought years. There were and are fewer highland drought refuges available, as administrators admitted in 1915: '[the Purko] have very little highland grazing at their disposal to fall back upon in periods of drought.'[33]

Much of the grassland was superb, including the valued *Themeda triandra* (red oat grass), the dominant species on the Mara plains. But tsetse and ticks rendered enormous areas out of bounds to herders. Robertshaw and Lamprey describe how 'the Maasai were confined, by tsetse infestation of the Mara Plains, to the [Lemek] area for the greater part of this [the twentieth] century'. By 1920 the shifting of different Maasai sections within western Narok, and changes in seasonal movement patterns, had 'left a vacuum on the Mara Plains into which tsetse bush expanded ... '. By 1940, tsetse had rendered around 900 square miles of grazing unusable, effectively creating the Maasai Mara game reserve.[34] Though wild ungulates were tolerant of trypanosomiasis, incoming domestic herds began succumbing to a disease then unknown in the cool highlands. Certainly up until 1918, it was still relatively unknown in their former environment.[35] In this area of the south the Maasai were exposed to two species of tsetse fly, *G. swynnertoni* and *G. pallidipes*, which mainly affected stock though both are also vectors of human sleeping sickness. Early reports indicated fly belts flanking the Mara River and a couple of other riverine sites. However, tsetse was not a major threat in this area before World War I; Maasai knew where the infested pockets were and avoided them. It was only later that tsetse extended its range, and became less manageable.[36]

Aneurin Lewis, in his 1930s work on ticks in the Southern Reserve (he also investigated tsetse), provided a useful if inexact breakdown of rangeland quality at this later date.[37] Out of the total area of nearly ten million acres, he claimed three million were arid. Several million more acres of otherwise fine grazing were unusable and avoided by the Maasai because of ECF, tsetse fly and ticks other than *R. appendiculatus*, which caused diseases such as sweating sickness in cattle. (Lewis and his team collected 30 species of tick, of which six were known to be vectors of pathogenic protozoan parasites.) What was left? Maybe at most there were three to four million acres of land for the exclusive all-year-round use by pastoralists. If this still sounds huge, it must be remembered that the newcomers were moving to the western extension, to join the thousands of Maasai already living there (up to 9000 Siria in Trans-Mara alone). European settlers, meanwhile, were believed to need an 'absolute minimum' of 10,000 acres apiece for an average-sized farm on Laikipia.[38] The KLC claim that each Maasai in the reserve had 200 acres per head begins to look less generous, certainly with regards to quality pasture for cattle; small stock are a different matter.[39]

The Maasai version of events

When asked about the moves, Purko elders insist on one thing. They say the British told the Maasai: '*Shomo Ngatet mikiwa ol-tikana*' or '*Shomo pii Ngatet, enakop o ol-tikana*'. The first means: 'Go to the south and may malaria/ECF kill you there!' The second variation translates as: 'Go completely to the south, land of malaria/ECF.' Interviewees made these claims before any question was asked about the incidence of disease in the two reserves. I spoke to several members of the Il-Terito age-set who were born in Entorror and took part in the moves as small children. One, Ole Mantira, insisted that he had personally heard the *askaris* tell the migrants to go south and die there as they forced the people forward, adding the distinction: 'I heard this with my ears. It was not something I was told.' The majority of my 64 interviewees made similar claims, but to quote a select few:

Ole Gilisho was told by the white man: 'Get out of Entorror because even us, we want to put our cattle and people here where there are no diseases' … The land was very suitable for the *in-kishu*. No diseases for both people and cattle, no diseases completely.

Muiya Ole Nchoe, then aged over 100, Lemek

Ole Gilisho tried to persuade his fellow Maasai to stay [at Entorror] because the place was sweet. [He said]: 'Let's don't leave, because there are no diseases for both livestock and people.' There was no ECF at Entorror, and for people there was just the normal flu and sometimes very, very rarely malaria.

> Leperes Ole Gilisho, son of Parsaloi, likely
> to be in his 70s, Lemek

There were absolutely no diseases here because Entorror was a very sweet place. There was no ECF because the land was not bitter. ECF is a very new disease in Maasailand.

> Tuarari Ole Sialala, 90s, Soysambu

What the police [attached to the move] were saying is: 'Go to Ngatet so you go and die of ECF and hunger.' They *knew* that Ngatet was dangerous because of diseases.

> Nteyo Ole Yiaile, 75, Ngoswani

Another son of Ole Gilisho, Mapelo, said of Entorror: 'There was no disease of cattle and animals, and no wild animals, so fewer [domesticated] animals died because of disease. If your cow gave birth to a calf you were sure of rearing it.' Though some insisted there were no stock diseases in Entorror, a few respondents conceded that there had been some. Leperes Ole Gilisho later contradicted his above statement and said there was rinderpest, foot and mouth disease and babesiosis. Olkitojo Ole Sananka recalled: 'There were only three stock diseases at Entorror: *ol-odua* [rinderpest], *ol-kipiei* [BPP] and *empuruo* [anthrax]. And for human beings there were no diseases, only *or-kirobi* [cold].' Ole Ndonyio said that the warriors who fought the British at Ololulunga and tried to return to Laikipia were driven by anger at the sickness that had struck down their people and herds in the south:

> It's the old memories that made them do that, because of the elders who were chased from Entorror … Because that land they were taken out from, people could not die of malaria and also it could not kill cattle. But when they came to this land, the *ol-tikana* was here and it was killing people and cows. And then the warriors said: 'The white people brought us here to kill us and our fathers! Because these our people are dying every day, cows are dying every day. Because they brought us to a bitter land. So we shall go [back]; we must go.' That was the anger that the warriors had, because of this *ol-tikana* killing people every day and killing cows, and it was not in that land they took us out of. So that is what made them rebel.

This explanation for the clash at Ololulunga in 1918 (see my thesis) was corroborated by Tarayia Ololoigero. He likened Entorror to a stolen cow:

> It is very painful for someone [to] whom you have not given your cow when they suddenly come to take it from you by force ... Another thing, they hated [the fact that] this side has no water, and they had left that place with water. This place has got ECF, and the former place did not have ECF. There was no ECF for cows, no malaria for people; there was neither a shortage of water nor shortage of rain, because they were highlands, and the lowlands were criss-crossed by rivers. So we cry for that land where food can grow and people don't get sick, or livestock ... Because we have come to a place [which] is bitter for both humans and livestock. Everybody just lives here by the grace of providence.

When talking about the qualities of land, some respondents accorded it human characteristics. Typically, they said it 'liked' or 'disliked' humans and animals. It even had the power to kill. Ole Teka quoted Ole Gilisho as allegedly saying of the British: 'The truth is, they chased me from my good land. They brought me to a land that hates cows and people ... it kills them.' Also constantly cited in interview, when elders were describing land and environment, were the binary opposites sweet (*sidai*) and bitter (*kidua*).[40] *Sidai* is widely applied both to the physical realm of well-being – plentiful pastures and supplies of permanent water, saltlicks and minerals, favourable climate, the absence of disease and other factors which lead to a proliferation of healthy cattle, plenty of milk and hence healthy people – and to social harmony associated with 'the good old days' at Entorror. 'Maasai society was *sidai* at that time because there was more respect ... people were staying in harmony,' said Lendani Ole Sialala. *Ol-odua*, on the other hand, is the opposite of prosperity and happiness. This state of being is described as bitter as the taste of bile. One sign of rinderpest is an enlarged gall-bladder. If rinderpest sweeps through the herds it also devastates society, leaving people with a bad taste in their mouths. Does a state of bitterness manifest in social disharmony? It would seem so.

Sondo Ole Sadera was born in Entorror and was told about the moves by his father. 'My father told me ... they never wanted to leave that place. Because that place called Entorror was good; there was no big drought – it used to rain every three months. And there were no diseases when cows met.' The British chased them out, he said, 'telling them: "Go to that land where malaria/ECF is!" ' And the directive proved true, because the *ol-tikana* was there, and it 'finished' cows and humans, as did BPP. Another illness, probably pneumonia, 'attacked the Il-Kitoip [age-set] when they were *il-aibartak* [newly circumcised], and it finished them'. His older brother Kurao was interviewed separately. Also born on Laikipia, he was small enough to be carried on an adult's back during the move, which indicates his likely age. 'We came to this place and a drought came and killed people here … and all the cows got finished.' When asked about the suitability of Ngatet he replied: 'This land had *ol-tikana* that killed people and also animals. Do you think we wanted a place that killed our *in-kishu*? But because they forced us to move, what are you going to do? Even when the white man was chasing us he was telling us: "Go and die of that *ol-tikana*!" So don't you think that they also knew that this land had *ol-tikana*? Then why were they telling our fathers like that? And it's true that we got that *ol-tikana*.' There are many other examples in my testimonies of this belief in a deliberate British action to exterminate the Maasai. A folk memory has evolved in which this idea is central.

Official views and interventions

The British knew very well, from their own surveys, what the quality of the Southern Reserve was before they sent the 'northern' Maasai there, although this was not admitted officially. Questions were also raised in parliament, at Leys's instigation, about its alleged unhealthiness, including deaths from anthrax. On 27 June 1912, for example, Harvey asked whether the government had investigated the healthiness of the reserve, and whether tsetse fly was to be found in any part of it. Harcourt said he had no reason to suppose the country was not a healthy one, but 'it would be impossible to give an absolute negative … without a prolonged investigation by experts. But the evidence before me seems to negative [*sic*] any idea of the fly being prevalent, and I need hardly say that I should not have sanctioned the removal of the Masai to the Southern Reserve if I had grounds for even suspecting that it was open to objection on that account.'[41] In fairness, he was being kept in the dark. The following year, government entomologist T. J. Anderson toured parts of the reserve to investigate the incidence of tsetse. He found that certain areas were unsuitable for stock, partly because of the presence of fly, but his report was not published until 1921.[42]

I have already sketched my hypothesis that British administrators' knowledge of the effective absence of ECF on Laikipia was a factor behind the second move. By 'effective' I mean that while it may have been present, genetic or acquired stock resistance (immunity may be too strong a term) rendered it relatively harmless. It is highly probable that Maasai herds on Laikipia were resistant to ECF, or a particular strain of it. The ability of indigenous Zebu and Zebu crosses to acquire immunity to ECF was known by 1910. Desmond Bristow, who has long experience of what he calls the 'scourge' of ECF in this area, believes that this was the case: 'They [the Maasai] could and did have resistant cattle. I don't believe there was no ECF in this area; I think their cattle were probably resistant to it and because of that they were not losing many to it.' Resistance would have been acquired through an attack in early calf-hood; some adult cattle then became carriers. The trouble would only have started when the cattle moved to the Southern Reserve, and met at least four new conditions: exposure to infected country *en route*, exposure to other strains, exposure to larger numbers of game (particularly buffalo, long-term carriers of the parasite), and higher concentrations of cattle in a more restricted area of grazing.

On the first and third points, the use of certain forest corridors as stock routes was very risky; it was common knowledge, says Bristow, that those leading out of Thomson's Falls were full of ticks. It is not clear if the eastern route used to take the Maasai south traversed some or part of the old connecting road between the two reserves, which had been closed in 1908 on veterinary orders after becoming tick infested. But if so, this corridor was also known to be rife with ticks.[43] They thrive in long coarse grass, which is sufficiently humid for egg-laying and the subsequent moulting of larva and nymph, and they multiply after the rains.[44] The forest corridors during and after the rains were ideal environments for tick populations, lying in thick shaded vegetation protected by overhanging trees. Also, a variant of ECF is bovine cerebral theileriosis (BCT) or Corridor Disease, so-called because it is picked up in forest corridors. It is caused by a very similar protozoan parasite, *T. parva lawrencei*, and transmitted to cattle from wild buffalo by the same vector tick as ECF.[45] Cattle resistant to or immunised against *T. parva parva* (the cause of classical ECF) often cannot withstand *T. parva lawrencei*. The strong possibility that while moving cattle were exposed to both parasites, bearing strains of ECF and bovine cerebral theileriosis to which they lacked resistance, must be factored in.[46] As for the risks arising from higher concentrations of cattle in a more restricted area, veterinary officer Francis Brandt wrote at the time of the move:

the Masai … with unlimited grazing, are accustomed immediately on the appearance of any disease to move their cattle to fresh grazing grounds, with, in the case of East Coast fever, a loss of only one or two head of cattle ... Infection in the shape of infected ticks is left behind ready to attack the next herd of cattle which pass. In this way, so long as the country is under-stocked, the losses are inappreciable, but in the event of an excess of cattle being grazed over a limited area an epidemic of East Coast fever would probably occur.[47]

Here was the recipe for disaster, and the vets were foretelling it before the Maasai went south. However, they did not see the Southern Reserve as 'limited' in size. Another factor was altitude, as the vets recognised very early on:

> In this country, with its varied altitudes ... it is easily understood that while the cattle indigenous to a district live and do well, remove them to other altitudes and a number of them will die. In 1899 a large number of transport cattle were removed from the Nandi Plateau to the low lying plains of Kavirondo. Many of these animals died from Texas Fever, and I pointed out at the time that dealers in livestock should remember that they had to undertake a certain amount of risk by removing their stock from one altitude to another.[48]

Another possibility, which may have 'disguised' the presence of ECF on Laikipia, is enzootic stability. According to this theory, ECF existed but there was little clinical manifestation of it. Maasai cattle acquired resistance, and disease would only occur after susceptible animals moved into the enzootic areas, or after enzootic areas were extended into contiguous but previously tick-free non-enzootic areas. The extension areas developed more grass and other ground cover after rainfall, new populations of the vector tick built up in the new vegetation, and epizootics of ECF broke out in susceptible cattle there. By this reckoning, the best thing to do with a tick-borne disease was nothing – except allow animals to develop natural immunity. The Maasai knew this then. The veterinary authorities also recognised the role of endemicity in the development of immunity.[49] Some vets now say this is the ideal strategy, though all farmers still dip in order to kill ticks.

Veterinary support

Veterinary interest in Maasai herds on Laikipia had been cursory before 1911. Collyer asked for better veterinary support in 1910. The Annual Report for 1912 was frank: 'Though a Veterinary Assistant has been in the District for three years the Veterinary Department first made a systematic investigation of disease amongst the Masai stock during the year under review. This was rendered necessary on account of the Masai move ....' Concern for the wellbeing of Maasai herds was not the priority; officials were anxious about settler stock since the Maasai were to be driven south across their farms. Two vets (Kennedy and Dixon) were engaged full-time on the 1912–13 move, together with the Chief Stock Inspector (Neave) and five junior stock inspectors. Incidentally, there is no mention in the veterinary reports of any attempt to dip cattle as a preventative measure before or while stock were being moved.

The main diseases among Maasai cattle on Laikipia between 1904 and 1912 were BPP and gastro-enteritis (more likely to have been rinderpest, see below), though an unnamed cattle disease restricted trade in the district during 1910. Rinderpest, BPP, black quarter (also known as quarter evil), *engamuni and m-benik*, described as 'possibly a form of East Coast Fever' were noted in 1911–12,[50] and redwater was mentioned in 1913. Rinderpest was thought to have long existed on Laikipia in an endemic form, striking down mostly young animals and leaving adult survivors immune. Some BPP was noted in 1906. This worsened by 1910–11 when it and gastro-enteritis were described as 'rampant', and an unnamed Indian veterinary assistant inoculated 'large numbers' of Maasai cattle against it. Collyer noted: 'A few years ago the Masai would have placed every obstacle in the way of inoculation but now they seem to like it.'[51] This refutes the idea that the Maasai resisted vaccination and other scientific practices at this stage of colonial contact. By 1912–13 the Maasai were said to have dealt effectively with BPP through self-imposed quarantine.

Gastro-enteritis 'swept through' the herds in 1909 and 1910, killing as many as 1000 head of cattle in some 'kraals', and an estimated 15,000 cattle died before the disease 'disappeared'. This figure included calves that had died or were born dead. However, the DC added: 'The Maasai persist in saying that the disease was rinderpest and they now consider their herds are immune from that disease.' Elsewhere, he noted that the Maasai viewed it as a *mild* form of rinderpest. It now seems likely that the Maasai were right – 'gastro-enteritis' was indeed rinderpest, in so far as rinderpest manifests as a gastro-enteritis, and was one of the commonest causes of gastro-enteritis in cattle at this time. They were also right about the link between exposure and immunity. 'A curious thing to be noted about this disease,' wrote Collyer of gastro-enteritis, 'is that [when?] it first appeared in 1909 it was fatal only to calves and very old stock, afterwards a new infection was introduced which was much more destructive, killing stock of all ages in large numbers; but the stock that had recovered from the first outbreak appeared to be immune from the second outbreak.'[52] The same principle applied to rinderpest. 'The majority of the Masai have had the disease nearly every year, and such people had no desire to escape it this year, being on the contrary quite willing to pay the necessary toll in young stock in order to have the survivors immune. It seems certain that in several instances they have purposely introduced the disease into their young stock with this object in view.'[53] It was the presence of fatal gastro-enteritis (rinderpest) in the herds at the end of April 1910 that 'suddenly' forced the authorities to delay the start of the move, according to the district record. But two other events also caused this delay: the Colonial Office had cabled Girouard with orders to halt the move until the Maasai agreed to do so, and Ole Gilisho had changed his mind about moving south.

Long after the Maasai had gone, Laikipia remained an officially ECF-free area. The first recorded outbreak in Rumuruti appears to have been in 1919–20, and originated in an animal from West Kenya.[54] In 1921, no cases were reported. In 1922, when there were still only 166 Europeans in the district, Laikipia was again declared clean. ECF broke out in December 1926 at Rumuruti among Somali stock that had moved from Delamere's Ngobit farm (on the road from Naro Moru to Ndaragua), and the township was placed in quarantine. By 1925, there were 36 reported outbreaks on Laikipia, which was one of three centres of infection in the country.[55] By July 1929, there had been 16 outbreaks of ECF during the year and it was described as 'a perpetual nuisance'. It remains so to this day. The evidence points to its likely introduction by settlers, who spread it via rail and road transport oxen.[56]

Besides the new conditions already mentioned, what greeted the Maasai on arrival in the Southern Reserve? Veterinary officer Stordy wrote that many cases of ECF had come to his notice, though he went on to say that large areas were 'sparsely infected' because of the Maasai habit of moving their animals away from infected grazing grounds, which had over time become 'automatically clean'.[57] His colleague Bill Kennedy, in a 1913 report on diseases in the reserve, said rinderpest was prevalent there before the move began. But it hit the incoming herds particularly hard because many younger cattle were susceptible. A 'serious epizootic' broke out, and by December rinderpest was 'very widespread' on the Mau, Melili, Loita Plains, in the Lemek area and on the Amala river. No cases of ECF were brought to his attention during his stay in the reserve, and he did not gather any historical information about it in the areas visited. He admitted: 'This does not preclude the possibility of East Coast fever existing in certain parts, however, as I did not have sufficient opportunity to carry out full investigations.' Large numbers of sheep had died of an unspecified disease on the Loita Plains between July and September, and there was some pleuro-pneumonia in goats.[58]

By the following year, a larger crisis was looming. World War I was fought in East Africa on the Maasai 'southern front' of the border with German East Africa. Though hostilities here were relatively minor, the war effort and preparations for it by both British and Germans involved major livestock movement – not only an increase in ox-drawn transport but also the mass movement of slaughter cattle acquired from the Maasai to feed the troops. This helped to spread ECF in Maasailand and elsewhere. Quarantine and restrictions on livestock movement were both eased during this period. Farmers even took their livestock into battle: 'Up to 80 per cent of the settler farmers joined the British army ... taking their ploughing and transport oxen to war with them.'[59] By 1917, all these factors led to outbreaks in Nakuru, the Limuru 'clean area', Naivasha and the Southern Reserve. According to the Annual Report: 'East Coast fever has spread considerably in the Masai Reserve during the year. This is largely, if not entirely, attributable to the movement of large herds of slaughter cattle purchased from all parts of this densely stocked area to meet the Military requirements.'[60]

By the time war ended, it was no longer possible to overlook the rising incidence of stock disease in the reserve. The Officer in Charge reported 'upwards of half a million deaths' in 1919–20 from tick-borne diseases, BPP and rinderpest. In January 1918, an outbreak of rinderpest in Narok District killed nearly all the calves in certain villages. In 1922, it killed between 60 and 100 per cent of all cattle in the Narok and Loitokitok Districts. Concerns about the spread of BPP led to the first veterinary laboratory being built in the reserve in 1918, to investigate BPP and other diseases such as foot and mouth and anthrax. Bill Kennedy, by now Acting Chief Veterinary Officer, travelled to Narok to discuss what could be done about BPP. He concluded that inoculating the three-quarters of a million or so Maasai cattle in the reserve would require a staff of at least eight veterinary officers and 50 stock inspectors. He only had 12 vets and 11 stock inspectors in the whole Protectorate.

Sandford gave a full and frank account of the diseases afflicting Maasai stock up to 1919. ECF and BPP were the two most serious diseases in the Southern Reserve; BPP had infected five 'villages' during the move, and these were quarantined. The whole reserve was placed in continuous quarantine from 1916 following major BPP outbreaks; quarantine restrictions on exports to Kikuyu and South Kavirondo Districts were not lifted until 1935. But ECF caused 'by far the highest mortality' of any disease in their herds. In August 1914, ECF had been confined to an area near Ngong, the Sotik border and Trans-Mara. Since then it had spread rapidly, and 'appears to be the only cattle disease which the Maasai really fear … '. Most significantly, he added: 'Masai cattle appeared to have bred a certain degree of immunity to the disease, but the Officer-in-Charge was inclined to think that the cattle which had come from Laikipia were less immune than those which had previously resided in the Southern Reserve.' He blamed the Maasai for not preventing it through dipping, by making use of the dip built at Ngong in 1914 at their own request. He put their reluctance down to superstition, slackness and listlessness.[61] However, my informants said that the Maasai initially resisted dipping because of the trauma it caused their beloved cattle, who were 'beaten severely to force them to jump'. A Maasai saying refers to this: 'We cannot drop cattle into a hole like warthogs.' It was also believed to weaken the cows' defence system, as well as 'hurt and drown their dear animals'.[62]

Were the veterinary authorities trying hard enough to treat and prevent Maasai stock disease in this early period? Leys thought not. He scathingly summed up the disparity between veterinary attention to white and black pastoralists in this period: 'The Veterinary Department professes to work for the benefit of European and African stock-owners alike. The claim is sheer nonsense. Nine-tenths of the Department's work consists of free preventative and curative treatment given to the property of Europeans, who own, according to official returns, only 5 per cent of the stock in the country.' Veterinary concerns about disease in native reserves, and their classification as 'dirty', did not match the numbers of vets assigned to tackling it.[63] However, that was hardly the fault of individual vets and veterinary researchers, many of whom were conscientious professionals and decent men with African sympathies, no doubt frustrated by government under-funding and official priorities. The priorities of staff at the research laboratories must be differentiated from those of the Veterinary Department, which was largely geared towards improving European production. But the fact that veterinary officers were employed almost exclusively in the European areas was freely admitted in Annual Reports between 1911 and 1924. For example, 'the energies of practically the whole of the veterinary staff have been concentrated on the prevention and, where possible, the eradication of stock diseases in the area occupied by

European farmers.'[64] The department did not intend to tackle disease in the reserves. The 1911 report stated: 'Eradication in the vast native reserves where East Coast fever is endemic is not to be thought of, even were it possible.' There is another curious anomaly here. Veterinary authorities reported that stock diseases were rife in the reserves, which therefore represented a serious menace to the stock industry. At the same time, they said that vets could only guess at what went on in the reserves, because they had a minimal presence there.[65] Without being there, and without testing, how could they be sure of the incidence of disease? The Acting Chief Native Commissioner raised the issue in the colony's Legislative Council in 1921: 'The native was far and away the greatest owner of stock in this country and up to then in the Reserves there had been nothing done for the natives from the veterinary point of view. It was difficult, indeed it was practically impossible to get veterinary officers to go there ....'[66]

The year before, the Acting Chief Veterinary Officer said quite bluntly that Africans were too stupid or stubborn to know what to do with vaccines and other modern inventions, so veterinary inputs were considered to be a waste of money in the reserves. For one example of the tone: 'The losses of stock throughout the native areas from preventable disease must be enormous but these would not be prevented at once simply by the provision of adequate veterinary staff and equipment. A considerable time would elapse in some instances before the conservative native could be convinced of the efficacy of the methods employed by the veterinary staff ....'[67] (At the same time, there were many references to white farmers ignoring or misusing veterinary measures such as dipping. And it was admitted that BPP had been spread not by ignorant Africans but by Europeans' use of oxen to carry goods from the railway at Gilgil to their farms.) As for anti-rinderpest measures in the reserves, it was admitted 'there was no policy'.[68] With ECF, the British were possibly mindful of leading bacteriologist Robert Koch's opinion that fencing, dipping and quarantine could not succeed in areas with large numbers of African-owned cattle. Therefore, with finances tight, they initially settled for quarantine only, confining Africans to areas that were by definition fenced 'reserve-oirs', or isolation wards. The African stock-keeper was caught in a Catch 22: his herds were diseased, he was offered little or no veterinary help to deal with the problem, and it was assumed that he would not know what to do with the help even if it were offered. So little or no preventative or curative help was given, and the government simply used continuous quarantine of the reserves to stop disease spreading. This prevented the Maasai from selling cattle and entering the market economy, which confirmed British criticism of their 'innate' conservativism.[69]

Human sickness

It is beyond the scope of this study to investigate patterns of human disease in the Southern Reserve. The earliest research into Maasai health there was undertaken by Orr and Gilks in 1926–29, but this was too narrowly focused to offer a reliable picture.[70] Suffice to say that the incoming Maasai, as well as their livestock, must have been susceptible to infections and diseases that were new to them. As McNeill writes: 'Clearly any change of habitat ... implies a substantial alteration in the sort of infections one is likely to encounter.' Also, many diseases have historically transferred from animal herds to human populations; measles is probably related to rinderpest, and smallpox to cowpox. Malaria and dengue fever 'may have been present from time immemorial, lying in wait for immigrants from more northerly climes among whom prior exposure had not built up any sort of natural resistance'. Many 'northern' Maasai may have had no prior exposure to human *ol-tikana*, hence the strong remembrance today of its deadly virulence. At the turn of the century the Hindes had noted, with regard to malaria, how many Maasai were 'always ill, and frequently die, on moving into infested districts' while those who had developed immunity could happily live in or pass through 'mosquito neighbourhoods'.[71] Most significantly, McNeill asserts that young adults are especially vulnerable to new infections: 'Sometimes new infections actually manifest their greatest virulence among young adults, owing, some doctors believe, to excessive vigour of this age-group's antibody reactions to the invading disease organism'.[72] This fits stories of how the Il-Kitoip, junior circumcision group of the Il-Tareto age-set, were allegedly 'finished' by disease, probably pneumonia, soon after they arrived in Ngatet.

The public health section of the Masai Annual Report for 1914–15 declared: 'The Masai seem to possess little stamina and quickly succumb to disease'. Mirroring the lack of veterinary support for sick livestock, the Maasai found, having succumbed to sickness, that there was no public health care system to help them. There was just one illiterate 'native dresser' at Narok, and no hospital, dispensary or medical officer in the entire reserve. By 1917, nothing had changed. 'Very little medical work is, therefore, possible ... Should a case of serious illness arise, it is almost certain that under existing conditions the patient would be cured or dead before medical assistance could arrive.' Dysentery and fever were said to be the main diseases. By 1923–24, the report noted that the Maasai were paying the highest rate of tax in the colony and were 'surely entitled' to medical help.

Leys did not investigate human health in the Southern Reserve; he had left the country by then. However, he knew a great deal about the kinds of infections that struck down migrants, because his work involved checking the health of migrant labourers. In discussing African sickness in general and British responses to it, he remarked:

> An erroneous idea prevails that Africans are immune to African diseases. This is quite untrue of the chief diseases. Nowhere in tropical Africa is the European death rate known to be so high as the African death rate. When large numbers of Meru from the slopes of Kenya were sent to work on the Mombasa Waterworks, they sickened and died just as Europeans or Chinese would have … the migration of such labourers not only exposes them to new infections, but results also in the migration of the diseases themselves. That is particularly the case with dysentery, epidemics of which are constantly breaking out in villages after labour gangs have returned to them.[73]

The forced migration to lower and hotter land, followed by the introduction of forced labour in road-building gangs and for other public works in the reserve, would have been important factors in the spread of disease among Maasai in this period.

Lewis's study of ticks

It was not until 1934, with the publication of entomologist Aneurin Lewis's pioneering study of ticks in the Southern Reserve, that a clear picture emerged of the extent of both tick infestation and other challenges to stock in Maasai country. This is a very rich piece of work, based on grassroots research in 1932–33, while Lewis was attached to the Veterinary Research Laboratory at Kabete near Nairobi. His findings confirm Maasai claims about the prevalence of ECF in the reserve, though he is less clear about why ECF broke out when and where it did. He also found evidence of other tick-borne diseases including Nairobi sheep disease, sweating sickness and heartwater. Lewis's description of the incidence and effects of ECF among Purko herds in western Narok leaves the reader in no doubt about the horrors of *ol-tikana* in that part of the reserve:

> The writer witnessed an outbreak of disease among adult cattle which swept away whole herds of hundreds of cattle. In one boma at Aitong, 300 adult beasts died within fourteen days of their return from the Mara ... dead bodies of sheep were strewn along the routes from Mara bridge to Engoregori. This is also true of the cattle. *Indeed the vultures, the hyaenas, jackals and other scavengers could not cope with the abandoned carcases.* Examinations of the dead animals and of numerous blood and gland smears, proved the cause of death to be East Coast fever.[74] (My italics.)

He claimed: 'Certain large areas of the Masai country are unsuitable for all stock; others are useful only for sheep, while still others are totally uninhabited by man or domestic beast.' The reasons for this were multiple: few permanent water sources, seasonal fluctuations in water supply which meant that certain areas such as the Loita Plains were only useable for part of the year, lack of grazing, the presence of tsetse fly and ticks, and fear of disease in areas where rinderpest and smallpox had previously decimated stock and humans. Where grazing was good, it was often rendered useless. Lewis witnessed mass starvation of domestic stock in the dry season. He was not surprised that Maasai were driven by stock starvation or poverty to 'trespass' outside the reserve.

He saw a link between weakened cattle and their greater susceptibility to ECF: 'It is true that, under natural conditions, there is a tendency for ill-conditioned, unhealthy and sick animals to become more liable to the attacks of ticks. Whether it is due to the condition of the beast and lack of resistance, or to the fact that such animals, by frequently resting often provide more time and opportunity for attack by ticks, it is difficult to say ....'[75] From this one can speculate that Maasai cattle weakened by the long march south from Laikipia arrived in a more susceptible state. The move also entailed frequent stops and starts, which would have increased the likelihood of attack by infected ticks *en route*. Modern veterinary opinion supports this idea.

Lewis plotted on a map the main dry-season migration routes of cattle in the reserve. Another map showed shaded areas where the chief agent of ECF, *R. appendiculatus*, thrived. Lewis noted that Maasai will venture into areas they normally avoid when faced by a critical lack of water and grazing, with one obvious result:

> these movements are towards and into areas infected with ECF. At Ngong it is known that Nairobi sheep disease exists also. In these

areas, *R. appendiculatus*, along with *A. variegatum*, is prevalent. When adverse conditions are at an end ... the Masai with the remainder of their stock wander back to their homes, away from the ECF infected areas. Obviously the stock infested in these areas with *R. appendiculatus* carry this species of tick – and others – to uninfected areas. Thus, bit by bit, new areas of the reserve become infested with this tick which may gradually become acclimatised, reproduce its kind and serve as a reservoir for the causal agent of the disease.[76]

Other factors which he suggested contributed to the spread of ECF included stock raiding outside the reserve, stock trading, stock exchange for bridewealth (particularly with the Kikuyu), mass movements of stock for ceremonial reasons such as circumcision, the practice of allowing Kikuyu to rent grazing in the reserve, the proximity of heavy infection in the Kikuyu and Ukamba reserves, and possibly intergrazing of Maasai stock with wild game such as wildebeest and zebra.

Lewis described Maasai knowledge of ECF, though he tended to dismiss their diagnoses of this and other tick-borne diseases as unreliable. They were aware of the differences between different species of tick (sing. *ol-masheri*, pl. *il-masher*), with ten variations on that term; for example, *ol-masheri onyukie* (*sic*, *nanyokie*) referred to all brown ticks. *Ol-masheri* was also used to refer to sweating sickness, another tick-borne ailment. He did not, however, discuss indigenous treatments. Merker had written much earlier of Maasai attempts to 'heal' ECF through cauterisation of the swollen lymph glands.[77] This tallies with current practice. Traditional treatments, which many people still use, tend to deal with the symptoms of disease. Hence, informants told me, the glands beneath the ear and/or neck are branded with a hot iron in the case of ECF, tobacco juice is inserted through the cow's nose in an attempt to 'stop the rotation attitude' of a cow suffering from circling disease (*ol-milo*), and the leaves of *Teclea nobilis* (*ol-gilai* in Maa) or *o-suguroi* are used to treat anaplasmosis.

Finally, Lewis attempted to answer the very interesting question of whether ECF had existed in the Rift Valley when the Maasai lived there – officially, it had only become infected in 1914, and was not known on Soysambu, Delamere's ranch, before 1927. He quoted Delamere, who believed ECF had been introduced to Maasailand from the coast many years before but 'had been got rid of by the Masai changing their grazing grounds'. Lewis surmised that when they had enjoyed the freedom to roam, the Maasai dealt with disease by quitting diseased pastures, only returning after months or years had elapsed. Ticks would have gorged themselves on the cattle and dropped off during the movement, or at various temporary camps where the Maasai rested at five- to ten-mile intervals: 'Ultimately all the ticks would drop off, deaths from the disease would cease and the people would be satisfied that they had left the fouled (or tick-infested) land. Movements of this kind resembled very closely the routine of the early "temperature camps" once adopted by veterinarians to control outbreaks of East Coast fever.' But confinement in reserves had put an end to traditional Maasai coping mechanisms.

> As a result of British administration the Masai were confined to reserves and their migrations were much restricted. The infestation of large tracts of their country by tsetse flies, the lack of water in other areas of the reserve, still further prohibited such movements as occurred prior to their limitation to reserves. Obviously this restriction prevented, to some extent, the old Masai custom of abandoning foul land and seeking clean pastures; it was accentuated when the stock increased and created a congestion of the reserve.[78]

Losses by other sections

George Goldfinch made intriguing references in the mid-1920s to losses suffered by other Maasai sections which had been forcibly moved to the Southern Reserve after 1911–13. He said the reserve 'has simply been a hotbed of disease for years', telling the Anti-Slavery Society 'nobody need believe any fairy tales about how nice for all the Masai to be together ...'. He made enquiries through Maasai contacts about the welfare of the Momonyot, 'Dorobo' and Uas Nkishu. Cattle belonging to these first two communities were dying of trypanosomiasis 'as where they were driven to has always been perfectly well known to contain fly ... The old Chief Legeshaur [?Legeshaun] came to see me about 10 days ago and told me they had then lost over 100 and others were sick.' Two years earlier, Goldfinch had been told by a son of Masikonde that the Uas Nkishu cattle were 'practically finished' by rinderpest and BPP, 'but what is perhaps worse is that he said lots of the people themselves were dead from disease'. This was described as 'Indegana of people ... This strictly speaking would mean human East Coast fever'. His informant said that people's complexions turned green, and they passed red urine. Goldfinch guessed that they were suffering from some form of malaria or tick fever:

> Anyhow I think it is a perfectly good example of what ... must happen if natives who have lived for generations in a healthy country are

non-resistant or mostly so to malaria or any other disease are moved into a district where those diseases are endemic ... This I think is a very bad case indeed ... Some people would perhaps say the Southern Masai reserve is a healthy country and so a good deal of it is but there are plenty of places that the original inhabitants avoid altogether or only go into at certain seasons. Strangers are of course forced into them.

A year later, in 1924, Goldfinch passed on more news of the Uas Nkishu. He had now been told that the disease they were losing their cattle to was in fact ECF, 'as the country they were forced into is an endemic area for that disease'.[79] In his 1913 report on Trans-Mara, nine years before some of this section had moved there from a northern reserve in Kisokon, Hemsted had told the CO he suspected that 'the tick which carries East Coast fever is fairly prevalent', but that the Siria cattle were immune.[80] There can be no doubt that the highest authorities were informed of these dangers at a very early stage.

To conclude, the available evidence indicates that there was indeed a sound scientific basis for Maasai claims that the immigrants faced severe pathological and environmental challenges on arrival in the Southern Reserve. It was a poor substitute for their former northern territory, and they lost the wide range of habitat necessary to their transhumant pastoralist practice – partly because other Maasai sections were already using the available resources, and would not necessarily share them. Their complaints did not simply amount to baseless 'griping', as Sandford and other colonial officials implied, but reflected very real and life-threatening concerns. With ECF, the evidence suggests that it was either unknown on Laikipia before 1911 or, more likely, that Maasai herds were resistant to the strains which existed there, and enzootic stability could be maintained. The cattle succumbed on moving to unfamiliar territory. Also, nomadic coping strategies involving move- ment away from tick-infested pastures kept ECF at bay or under control; therefore it is simply not remembered as having caused a problem. The stress of the move is likely to have weakened the cattle and predisposed them to inter-current disease. Losses were exacerbated by the appalling lack of veterinary support in those early years.

At the same time, climate change must be factored in; as the twentieth century advanced, this must have led to more droughts and triggered other eco-system change which made populations more vulnerable. And the denial by some elders that *any* stock diseases existed on Laikipia when the Maasai lived there is patently untrue. The veterinary and dis- trict records detail the incidence of sickness in the 1900s, including the fact that 60,200 head of cattle were infected by rinderpest in October 1912, which presented a major problem for the vets charged with managing the move. One may conjecture that Maasai simply mean diseases new to them since they moved south, such as 'new' strains of ECF. And that the folk memory of late nineteenth-century epidemics, the disaster called *emutai*, is so frightful that it has eclipsed smaller outbreaks of disease in the period between *emutai* and the moves. By comparison, any lesser calamities were relatively easy to dismiss. Also, veterinary records suggest that the birth rate greatly exceeded the death rate to disease in the reserves; stock numbers were so high that losses were easily absorbed, herds recovered fairly quickly, and therefore losses could be 'forgotten'.

Part III
Interpretations

6
Blood Oaths, Boundaries and Brothers

> I know about the oath, with its agreement 'please don't swallow
> me because I can't resist you'. Ole Gilisho decided [to do] this,
> because he did not want the battle.
>
> Iloju Ole Kariankei, grandson of Ole Gilisho

Before Europeans arrived on the scene, the Maasai were the acknowledged 'lords of East Africa', driving all before them. So why did they meekly allow white settlers to take their best lands, and why did they not violently resist the forced moves? They were not in a strong position to do so, as mentioned at the start. But the answer may also partly lie in a blood-brotherhood oath or peace treaty, said to have been made by leading settlers and Maasai representatives beneath an ancient fig tree on Lord Delamere's Soysambu ranch in the Rift Valley, sometime before 1911. Go there today, guided by Maasai farm workers, and one finds an atmospheric spot – a natural conference site of wood and rock, bounded by a stream and an orange grove planted by the current Baron's son and heir, Tom Cholmondeley. What might have happened here, and why do certain Maasai still talk about it a century later, even if the British have long forgotten?

There is virtually nothing about this in the literature. Elspeth Huxley mentioned it briefly, but gave no details. She described how Lord Delamere (Hugh Cholmondeley, the third Baron) went to Ololulunga to mediate between the British and the Maasai after the 1918 battle, where he met Ole Gilisho and reported his reaction to the tragic events. Quoting Delamere, she wrote: ' "Legalishu [*sic*] himself is anxious that the *moran* should come into line but he is very upset about the accident – the shooting of women and cattle. He says he made blood-brotherhood with the Government years ago, and that meant that you and he were of

135

one blood and would not spill each other's, but that this accident has upset that ... ." '[1] Gerald Hanley also briefly mentioned an oath, suggesting that Maasai leaders and British officials took one after the Kedong massacre of 1895. He quoted Collyer's interpreter Ole Kirtela as saying, 'oaths were taken on both sides swearing that neither side would look for revenge and that they would live peacefully'.[2]

Maasai elders talk about the blood-brotherhood ceremony as if it happened yesterday. They bring it up without being asked, within minutes of starting to talk about the moves. The simplest description was given by one of my interviewees, Olokimolol: 'The oath took place to stop the violence between the two groups – the whites and the Maasai. So the whites became our brothers and we became their brothers.' Forty-eight out of 64 interviewees said they knew about a Maasai–British blood oath, and many were able to describe it. There was only one outright denial that such a thing had taken place, and that was later modified. My first interviewee, Daudi Ole Teka, first raised the subject. Until then, I had not considered asking any questions about an oath since I had never heard of it; the questionnaire was subsequently modified. Even then, some elders continued to mention it spontaneously, early in the interview, before the 'oath questions' were put. It was clear from what people said, and their impassioned tone, that the alleged oath was a significant feature of the colonial relationship and informants' understanding of it, and that elders set more store by it than any formal agreement with the British. Indeed, many respondents had never even heard of the formal Agreements of 1904 and 1911, or did not know what they contained. The oath was often spoken of in the same breath as the boundaries that divided the British and the Maasai after 1911. Said Ole Teka:

> When the British came they saw that the Maasai were not good, so they wanted to cut the Maasai so that they can share that blood [he indicated cutting a wrist]. They wanted to cut the Maasai to drink the blood so that they could put the boundary where it is now, and that is how we, the Maasai, were brought on the other side of the boundary, because of Ole Mbatian's agreement with the white men.[3]

The oath itself seems to have symbolised a kind of boundary to the Maasai, both time-wise and spatially, since it is spoken of today as a major watershed in their colonial experience. What exactly happened? The majority of elders claimed that Ole Gilisho represented the Maasai at the ceremony, and four believed that Olonana had been involved, too.

However, the prophet's grandson, Olkisonkoi Mako (aka Oloruma), said he did not know anything about a blood oath. A majority claimed that the European participants included Lord Delamere and Gilbert Colvile, known to the Maasai as Nasoore.[4] A few informants suggested that other leading settlers in their coterie also took part, including Cole (it was not clear which brother, Galbraith or Berkeley), and settlers known by the nicknames Swara and Kakaangi. I later asked Arthur Cole, son of Galbraith, whether his father or uncle had ever mentioned such a thing. He replied: 'Wouldn't demean himself [*sic*]. Masai despised blood oaths'.[5]

Not everyone said the oath took place on the Delamere ranch; most people did not know for sure. Many said vaguely that it happened in Entorror, their former northern territory. As for when, it is almost impossible to know that either, since elderly Maasai do not think in terms of calendar years but warrior age-sets. The consensus is that it took place sometime between the arrival of Delamere in BEA (he first visited in 1897 and settled in 1903) and the second Maasai move. Some said it coincided with the making of Olonana's boundary, which is believed to run along the eastern Rift Valley escarpment above the town of Mai Mahiu. There are many stories of how Olonana ritually created a boundary between the white and Maasai communities by waving a *rungu* (stick) to demarcate the land. Maasai believe that this can still be seen today, since no vegetation has grown along the boundary line. Mure Ole Kamaamia declared:

> That was the day the blood oath was taken. The Maasai were told, if you come beyond this boundary it will be a different case. The whites should not cross the boundary and the blacks also. During that time of the boundary, the black and the white, each was cut on the tip of the finger for the blood, and the *ol-aiguenani* of the Maasai and the white person, one of them also sucked the blood. Delamere was among the people who was making the boundary.

The majority of informants described the ceremony in this way: the Maasai and British participants cut their own wrist or forearm, and each sucked the blood of the other. A couple of elders, Olkitojo Ole Sananka and Ole Kamaamia, said the participants pierced an index finger under the nail, and drew blood that way. Some claimed a large meat feast was held to mark the occasion. Olochani Ole Karbolo said this took place on the Delamere farm.

> A big oxen was slaughtered, the meat was roasted, and then on that particular day people were not given anything to drink but fresh

water ... mixed with milk. The blood was taken by both parties, the whites and the Maasai, signifying that there was no quarrel between both communities.

Ole Karbolo also said that Delamere was so close to the Maasai that he shared *en-kiyieu* with Ole Gilisho, a ceremonial 'sharing of the brisket' which establishes deep friendship between people.[6]

That [*en-kiyieu*] means we only stop our quarrel between the whites and the Maasai if we share that one, that thing called *en-kiyieu*, and Delamere accepted that also ... So this kind of oath taken was a trick used by the Maasai [when] they saw that they could not face the whites because of the [superior] weapons. Because no one can face the white man's firearms, the Maasai said: 'Let us be in peace'. So the Maasai were very happy that they were at peace with the whites.

Aspects of this version of the story were corroborated by Lendani Ole Sialala, retired livestock headman to the current Lord Delamere, who worked for the Delameres from 1964 until his retirement in 2000. His family was among those that returned from the Southern Reserve to find work on white farms, after they lost their cattle to diseases in western Narok. He gave a second reason for the oath, which sounds highly plausible in the context of the settler–worker relationship (see next chapter), and adds weight to the argument that Delamere took part in the ceremony.

The meaning of the blood oath was to make a relationship between the whites and the Purko Maasai, and the other meaning was so that the Purko would not steal Delamere's livestock, because he had decided now to come and stay with them, and so they were making that relationship. Delamere was playing a trick so that the Maasai wouldn't steal from him ... The oath took place within Delamere's farm, under a tree called *oreteti*,[7] next to a place called Lanet that is just within the farm. At that time they slaughtered an ox, and then they removed the heart of the ox and brisket. And then Lord Delamere had to bite the brisket four times and the heart four times, and then hand it over to the other Maasai elders. There was one called Namantile and another one called Lemorinke Ole Oyie, and the *ol-aiguenani* Ole Gilisho was also there.

Ntomoilel Ole Meitaya remembered: 'Nasoore was always eating meat in the bush together with the Maasai warriors, and also Lord Delamere was always with the Maasai elders. They were the ones who came with Ole Gilisho; they came and took a blood oath ... that united them to be brothers. So when the blood oath took place, there was peace between the whites and the Maasai. Then they came and divided the land and Ole Gilisho and Olonana were the ones who were dividing [it].'

The only outright denial came from Swahili Musungui, former gunman and interpreter to Colvile.[8] He claimed to be a member of the Il-Damat section, but other Maasai in the neighbourhood insisted he was 'Dorobo'. When asked if he knew anything about an alleged oath, he replied crossly: 'Don't listen, nothing like that. Nothing completely! Let nobody cheat you that there was something of that sort. I know very well because it was Delamere who was said to have shared the blood oath, but there was nothing.' However, he immediately qualified this by suggesting that a pact was agreed over boundaries. 'There are many places in Kenya where the Maasai are not living now, but they have Maasai names. For example, Ongatapus, Lariak, Gilgil, Ormuteta. *So that was the oath.* The Maasai told Delamere: "You have taken a very big land, so please don't follow us to this place [Ngatet]." So they came to an agreement – "this is the boundary", and there was not any argument any more.' It is possible that Swahili denounced the oath because he believed that only he, not the Maasai as a whole, forged a special relationship with Colvile and, by association, with Delamere. This will be explained in the next chapter. He may also have seen himself as a moderniser, who had no truck with archaic practices. He denounced prophets, too, saying: '*Laibons* are liars, we don't believe in them. Personally, I don't need anything to do with *laibons*. If they come I tell them: "Fuck off!" ' He lapsed into English with gusto.

Why was blood-brotherhood made? The short answer is that it was timely and necessary. The Maasai feared that they would be 'finished' by British weaponry, and that the British might turn their guns on the warriors to stop them from raiding. Sondo Ole Sadera said: 'They took the oath because the white man was telling them not to go raiding. And then Ole Gilisho told them: "Let's take an oath so that you can't fight me, because I can't fight you." ' According to informant Olololoigero, the age-set spokesman is also believed to have said: 'Let's take this blood oath so from henceforth everyone will be on his side of the boundary.'

Tragically for them if an oath was made, the Maasai chose unrepresentative whites with whom to strike such a deal. It is easy to imagine how they mistook settlers for government officials. For one thing, mutual confusion would have arisen around translation of the word *ol-aiguenani* (age-set spokesman, often wrongly translated then and now as 'chief'). In explaining his lordly position, Delamere may well have described himself as an age-set spokesman or chief of the Europeans. Informants clearly believed that Delamere (or Colvile) led this community. Mure Ole Kamaamia's description is typical: 'Delamere was the overall leader of the whites.' Or Ole Kimolol's: 'Nasoore was the *ol-aiguenani* of the whites in Entorror. He captured the land.' Delamere was an unofficial member of the Legislative Council from its inauguration in 1907, later an elected one, and remained a member on and off all his life.[9] But much more visibly to Africans, he appeared alongside government officials at important meetings with the Maasai. For example, he was at the key meeting on 24 February 1910 at Kiserian, at which Olonana, Masikonde and Ole Gilisho reportedly agreed to move south.[10] In 1918 he represented the government at the Ololulunga mediation, accompanied on this expedition by his brother-in-law Galbraith Cole, according to Cole's wife Eleanor. As she grandly described it: 'Galbraith and Delamere were asked by the Government to go into Masai land and explain some Government policy to them.'[11] Key meetings aside, the seat of government was a million miles away to most Maasai; their white settler neighbours were the immediate face of colonialism. Ole Gilisho's son Mapelu said:

> That white man [Delamere] came to Narok, and he was a great friend of Ole Gilisho, so when the Maasai saw him they thought that he was also a government representative. When they saw a white person they did not differentiate between an official and the settlers. So Lord Delamere has taken that advantage of acquiring a lot of land the other side of Mau, cheating the Maasai, using Ole Gilisho, also posing as a government official.

Retired headmaster Joseph Ole Karia, former chairman of the UN's African Indigenous Peoples' Forum, and later its representative for East Africa, added: 'The Maasai were not reaching the official people like Charles Eliot. The people they were reaching were the settlers who were living around the Maasai, settlers like Lord Delamere ... But being white, they [the Maasai] never knew the difference. Being white, you are always the government. It must have been a very special thing, because the Maasai have never made a blood oath with anybody else.' This last point was corroborated by others.

Gilbert Colvile's closest surviving relative, his adopted daughter Deborah, did not know anything about his alleged involvement in an oathing ceremony and believed it to be unlikely.[12] The current

Lord Delamere, who worked for Colvile as a young man and described him as an 'amiable pirate', had heard of the blood-brotherhood but could not remember where or when. He believes it did take place, and that his grandfather would have taken it seriously. Delamere is likely to have read Lord Lugard's account of his early years of empire making in the region, in which he described making blood-brotherhood with a succession of East African chiefs on behalf of the IBEAC.[13] The treaty he agreed with Musoga chief Mbekirwa, on the day they became blood-brothers, specifically declared that neither side would allow their followers to steal from or molest the other.[14] This was good news for cattle-keepers; Delamere may well have thought it advisable to take out similar insurance, given the Maasai reputation for cattle snatching, not that this was preventative as things turned out. To speculate further, he was drawn from a social class that had parallel rites of passage and ceremonies associated with 'age-set' affirmation – the ritual 'blooding' of youngsters on their first fox hunt, the initiation rituals for new boys at Eton and other leading public schools, and archaic all-male practices and slang peculiar to Eton.[15] Therefore a ceremony of this sort, involving men only and a measure of secrecy, would not have seemed alien. Most importantly, both men may well have been flattered that the notoriously hostile but 'noble' Maasai had chosen to bond with them and not others.

The Maasai also had good reason to make peace with their new neighbours and clarify the boundaries between the two communities. They were particularly anxious, on having agreed to go south, that the British should not 'follow them' there. The fact that they did so, and that British administration intensified if anything after the second move, came as a complete shock to them, according to my informants. Those who signed the Agreements, and may also have made an oath, are said to have assumed that each community would keep to its own agreed territory without interference from the other. They believed that that *was* the agreement. For example, Reverend Ole Kasura said: '[They thought] the oath represented the agreement that we are not going to refuse to give you this land. And if we give it to the British, we take the oath that you are not going to follow us to where we shall go ... .' Likewise, Ole Kimiri said: 'The oath itself signified that we the Maasai, we shall go and never come back again to Entorror, and the white people will never go back to Ngatet to go and disturb the Maasai.' These are old and often repeated refrains. The evidence which Maasai gave to the Kenya Land Commission in 1932 also suggests that they believed the European visitors were temporary, as was their occupation of Maasai

land. 'They [the Europeans] said "Give us a place to dwell in your country", and we said "Very good". According to our customs and knowledge
people who came into a country only stayed three or four years then
went away,' witness Arthur Christopher Tameno told the KLC. 'They
made an agreement with us for one part of the country.'[16] Kikuyu evidence indicated similar beliefs – that Europeans were initially regarded
as temporary occupiers who would go away.

As for whether the Maasai now believe that the oath has been broken,
some elders say the warriors broke the terms of the oath, in so far as they
continued to raid and spill blood. However, the majority consensus is
that the oath is intact and the British and the Maasai are still blood
brothers: 'We have become one with the British. Right up till now, we
are still one with the same people,' declared Ololoigero. 'It is only the
warriors who have erred and brought some trouble. The elders did not
want them to fight … more so when they have taken the oath.'

Blood-brotherhood in the literature

Contrary to modern day claims, descriptions exist of Maasai blood-
brotherhood with other peoples in the past. This is how Hollis described
it: 'A Masai elder would sit down with one of the elders of the savages
[people of another tribe]; each of them would then cut his left arm, and
after dipping in the blood some meat of a bullock which was killed on
the spot, would eat it'.[17] However, he differentiated this kind of ceremony
from that held to make 'solemn peace'; after making blood-brotherhood,
he claimed, the Maasai did not keep the peace. When they wanted
peace, they exchanged cattle with the enemy and 'the enemy's child is
suckled at the breast of the Masai woman' and vice versa. He claimed
that the Maasai made peace in this way in 1883 with the Lumbwa
(Kipsigis) at the Ford of Sangaruna, on the Pangani river, and referred to
descriptions in Hobley and Johnston of a similar ceremony after a war
between the two parties.[18]

McClure described three different forms of oath that the Maasai
considered to be binding. One was the 'cementing of a treaty by the for-
mation of a blood-brotherhood', but he regarded this as being of Swahili
origin and rarely used. Another involved 'the suckling of the children of
the opposing tribe at the breasts of Masai women and vice versa', and
a third the killing of a bullock, sheep or goat and the eating of portions
of the breast and other parts of the animal. Later, the participants wore a
piece of the animal's hide on their wrists, like a bracelet. The breaking of
this oath was supposed to bring sickness or even death on the offender.[19]

Macdonald briefly mentioned that the Maasai and the Kamba 'have made a sort of treaty [which] ... is on the whole wonderfully binding on the Masai'.[20] Francis Hall described a peace-making ceremony between Kikuyu and Maasai who had sought shelter at Fort Smith in 1894, which echoes McClure's account. In his diary, he noted on 31 January that year: 'All the friendly Wakikuyu assembled to eat a sheep with the Masai and generally perform the ceremony of making friends. The Masai then each wear a piece of the skin round their wrists as a token and are now free to walk about in the immediate vicinity though I have warned them not to go far.'[21]

Tegnaeus has described the practice of blood-brotherhood in many parts of world.[22] He listed some 42 ethnic groups in East Africa and the lacustrine region that practised some form of it, and he included a useful overview of the early literature on Maasai blood-brotherhood. He referred to it as 'an expression of friendship between both tribes and individuals', but said the Maasai did not take it too seriously. He cited references in Hollis, Merker, Neumann and Weiss, all of whom wrote before 1910. Merker believed the custom was not native to the Maasai, and Hollis echoed this: the fact that they only made blood-brotherhood when they did not seek 'solemn peace' suggested that it was a borrowed ritual. Some Maa-speakers only made it when it was politically expedient. 'The Masai (settled among the Djaga), have no blood-brotherhood, for which reason they are held in bad repute among the Djaga as violators of the blood-brotherhood. The ceremony itself was of no consequence to them, when, for political reasons, they submitted to it.'[23] This is presumably a reference to the Arusha Maasai, living alongside the Chagga people in what is now Tanzania.

It is best to quote Merker directly. He described two varieties of blood-brotherhood, called *ol-momai*.[24] Both were peace-making ceremonies, one forged between men and the other made by women. Before the first ceremony, some old men would visit the 'tribe' with whom they had been at war, carrying a bunch of grass and taking with them a sheep wearing a necklace of blue or green beads. These were peace tokens (*intokitin osotua*). The following ceremony would take place between elders:

Both parties sit down under a shady tree near the village, surrounded by the warriors, a strange elder opposite to each of the delegates. Then each man makes a small cut in the left forearm of the man opposite him, wipes off the blood that oozes from it several times on a piece of half-roasted meat, which he then eats. During this ceremony the parties concerned swear eternal peace. The meat used comes from a newly-slaughtered animal, usually an ox, less often a goat or sheep.

Merker claimed that the Maasai had deceived Europeans several times with this ceremony, as well as fellow Africans. He called it 'the false use of blood-brotherhood'.

> The author believes that there can be no possibility of concluding a lasting friendship between Masai and Europeans, both as a result of eight years' experience, and from information given him by a large number of Masai who, after a long personal acquaintance, trusted him enough to speak openly on this point. In their opinion, such a friendship ... merely has the result that individuals in the service of the European concerned are hospitably received in Masai villages. It cannot be called a lasting peace, any more than it is an inducement to the Masai to observe European law and order.

He went on to describe the entirely different strategy that was used when the Maasai were 'in earnest about a peace'. He gave an example of their peace making with the Kahe, calling this ceremony *ertana etabashage* or 'the nursing of the exchanged children'. It took place between two women and their unweaned children, in the presence of witnesses. Each woman would swap babies, and put the other's child to her breast.

Tegnaeus said Weiss gave a very similar account of Maasai blood-brotherhood, and followed Merker in emphasising the comparative seriousness of the suckling ceremony. Neumann, traveling in the 1890s, made blood-brotherhood with the 'Wakwavi', formerly mistakenly described as an agricultural branch of the Maasai. His book on elephant hunting has photographs of his blood-brother, Ndaminuki, and of his companion Dr Kolb taking part in the ceremony. Neumann joked: 'We had no difficulties with the natives, and entered into friendly relations by "eating blood" with most of the tribes we passed through – indeed I and my African brothers form quite a good-sized family altogether ... .' He would never have found so many elephants without having entered 'the bond' with Africans who showed him their 'haunts.'[25] Hollis claimed that the neighbouring Nandi did not use blood-brotherhood in the old days; it was a custom introduced by coastal traders. They never considered it binding, hence Peter West's fate (see below). But Huntingford claimed the Nandi had two borrowed forms of the ritual – one taken from the Maasai, the other introduced by Swahili traders.

As for the Kikuyu, according to Routledge their blood-brotherhood ceremony differed in essential details from that ascribed to the Maasai by Hollis, Merker and my own informants. Again, it is preferable to cite Routledge directly because Tegnaeus's précis is confusing. Incisions were not made in forearms but at the end of the sternum and on the forehead. A sheep was killed, the heart removed and roasted, and the human blood placed in the sheep's heart. This was then cut in two, and each participant ate one half. Routledge also described a second ceremony, involving an exchange of animal blood, to mark 'the reception of a [Kikuyu] stranger' into a new district. This involved mingling the blood of two slaughtered animals; both participants would then wear a leather bracelet made from the hide of the other's sheep or ox.[26] In other accounts, Arkell-Hardwick described making blood-brotherhood with a Kikuyu chief. It was called *muma*, the plural of the Maasai term, *ol-mumai*. Cuts were made in the chests of both people, a roasted sheep's liver was cut into small pieces, dipped in the human blood, exchanged and eaten. This was repeated three times.[27] Von Höhnel, travelling with Teleki among the Emba 'in the Kikuyu region', also called it *muma* and said it took place almost every day.[28] However, Leakey later dismissed Von Höhnel's description of blood-brotherhood as totally foreign to the Kikuyu. Since they did not consider these oaths binding, they did not keep them.[29]

Leakey suggested that the early European travellers to East Africa would have faced fewer problems if they had either made use of the ceremony of mutual adoption practised by the Kikuyu and the 'Dorobo', or a peace-making ceremony used by the Kikuyu and the Maasai. The first was intended to bind two families after a 'Dorobo' had sold land to a Kikuyu. Wrists were bound with rawhide straps, in a description that echoes Routledge's, and the sons of both men ate some roasted breast of ram before taking the same oaths as their fathers. An ox was also slaughtered and eaten. This was followed by a 'ceremony of showing the boundaries', when the 'Dorobo' showed the Kikuyu the key landmarks on the boundary of the land concerned – such as large trees, old game pits, rocks and streams. Many of these were anointed with a ram's stomach contents. The peace-making ceremony with the Maasai was always initiated by them, and did not involve blood letting. The Maasai participants simply had to bring tokens of peace – some butterfat, an arrow for bleeding animals, a branding iron, a pair of sandals, a bunch of grass and a gourd of milk. They also brought a young girl covered in butterfat and red ochre, accompanied by her mother, who were themselves a sign of peace. The Kikuyu would produce similar tokens. The fat of a ram was sprinkled over the Maasai objects, the Maasai chewed some grass and spat it over the Kikuyu elders and their objects, and both parties then ate the meat of the ram together.[30]

In other sources, British administrator Francis Hall described a peace ritual performed by the Kikuyu and the 'Chamvu' [illegible] people at Fort Smith. It involved the grisly torture of a goat. 'It is a beastly cruel operation and typical of the nature of the Wakikuyu,' wrote Hall, describing the ceremony as 'their usual one'.[31] In recent times, Waller has referred to a 'unique' peace ceremony that ended the Iloikop Wars with the surrender of Senteu to his half-brother Olonana in 1902. A deputation came bearing peace tokens including axes, headless arrows and necklaces made of black cowrie shells. There appears to have been no exchange of blood.[32]

Whatever the specific practices, which varied from one ethnic group to another, it is clear that an exchange of blood generally signified formal peace making in this region. More generally, as Luise White points out, 'blood ties, biological or socially constructed, are ways of creating bonds of intimacy and loyalty'.[33] Whether or not the ceremony was binding, and whatever its origin, there were cross-cutting similarities in the way it was done.

Treaty making

European treaty making with African chiefs in the nineteenth century often included a blood-brotherhood ceremony, according to contemporary accounts. In the British sphere of East Africa, these treaties were largely made on behalf of the IBEAC, and were the basis upon which the Company's claims to territory rested.[34] Henry Morton Stanley made blood-brotherhood as part of his treaty making with various chiefs while on his way to the coast with the Emin Pasha Relief Expedition. The most eminent was Ntale, King of Ankole, though Stanley performed the 1889 ceremony with Ntale through his 'son' Buchunku or Uchunku. (Gray claims he was not the son but simply a member of the royal clan.[35]) Ntale went on to make a treaty with Lugard in 1891, through his son Birinzi, and again in 1894 with Major Cunningham acting on behalf of the British government. The first of these definitely included blood-brotherhood. Jack Foaker, government representative at Guasa Masa, made blood-brotherhood with the Nandi in 1896. Matson described it as a 'routine practice' in Nandi country.[36] Frederick Jackson made it with the Lumbwa (Kipsigis) and others, but took a different tack with the Maasai; though he met unnamed Maasai prophets on the Kinangop and at Naivasha, including members of Mbatiany's family, he gave out blank treaty forms as 'medicine' but did not make blood-brotherhood.[37] Freelance European explorers and traders also used blood-brotherhood to cement new friendships. Von der Decken mentions being greeted as 'our friend and blood-brother' by the Sultan of Dafeta [sic] and his advisers.[38] Peter West, trading partner of the infamous Andrew Dick, is one example of a trader who indulged in the practice – Ternan mentions that he made blood-brotherhood with Nandi chiefs. But it did not save his neck; in July 1895 West was killed by the Nandi in revenge for Dick's murder of two Nandi.[39]

The fullest description comes from Lugard. He made blood-brotherhood on at least six occasions between 1890 and 1891 on behalf of the IBEAC: with Birinzi (or Bireri), son of King Ntale of Ankole (1891); with Wakoli, chief of Usoga or Musoga (1890); with chief Mbekirwa or Mubikirwa, chief of Buyende (1890); with several Kamba chiefs (date unclear); with Waiyaki and other Kikuyu leaders (1890); and with chiefs Katonzi and Mugenyi, respectively near the Kavalli plateau and Semliki (1891).[40] When he set out, he did not know that Stanley had already secured treaties with some of these leaders. On meeting Ntale's people, 'I greatly pleased them by consenting to go through the full ceremony according to their own rites.' The ceremony was different with Wakoli: 'a coffee-berry is used, and we rub each other's shoulders with the right hand, and then shake hands vehemently ... .' With Mbekirwa, he again made both blood-brotherhood and a treaty. The Kamba 'were most friendly and begged me to make blood-brotherhood with several of the chiefs, who had not hitherto made alliance with the Company. I did so, and we parted on the best of terms'.[41] With the Kikuyu: 'I made treaties with Eiyeki [Waiyaki] and several other chiefs, who came from considerable distances to perform the ceremony of blood brotherhood ... The fact is that, though I was provided with "treaty forms", I did not see my way to using them.' He believed them to be dishonourable because he could not guarantee that the IBEAC could provide protection, while asking a chief to cede all rights of rule was unfair. Also, 'the nature of a written compact was wholly beyond the comprehension of these savage tribes'. Hence he resorted to the pre-existing 'solemn form of compact for friendship' – blood-brotherhood. It allowed him 'to say just so much, and no more, as seemed a fair and honest bargain'. However, he still wrote down their 'mutual undertakings' and chiefs set their marks on the resulting documents, which were sent to the Foreign Office for approval and registration.[42]

Lugard described the method of making brotherhood, which varied slightly from tribe to tribe: 'each of us cuts our forearm till the blood flows; the arms are then rubbed together to mix the blood, and two small pieces of meat are supposed to be touched with the blood: he [the chief] eats the piece which has my blood on it off the palm of my right hand, and I eat the piece which has his blood on it from his palm ... Sometimes salt or a coffee-berry (in Uganda and Unyoro) is substituted for the meat. Sometimes the incision for bleeding is made elsewhere than on the arm.'[43] Great speech-making would follow, and pledges were sworn upon weapons. Lugard gave a paper on his travels in the Uganda region to the Royal Geographical Society in November 1892, in which he elaborated on this and introduced his do-it-yourself guide to treaty making in Africa. This appeared in the first issue of *The Geographical Journal* the following January. It included a graphic description of how a blood-brotherhood ceremony should be conducted by representatives of the Crown, and why it was necessary to use a pre-existing ritual:

> We hold a written bond 'in black and white' to be a sacred thing, binding in a peculiar way on those who deliberately sign it. But this is a civilised idea, foreign to and in no way understood by a savage. There exists in Africa, however, a parallel institution, and when I learnt its significance it seemed to me that I had found the nearest equivalent possible to our idea of a contract. This is the ceremony of blood brotherhood ... .[44]

All these accounts indicate many regional precedents for ceremonies from which the Maasai may have borrowed, and vice versa. Inter-ethnic borrowing was and is common, and includes language (loanwords), institutions, adornment, clothing, arms and shields, adoption of ritual leaders/prophets, elements of circumcision and warrior rituals, and intermarriage. There is therefore no reason to suppose that this practice was not also borrowed from neighbours such as the Kikuyu, and adapted for use on select occasions. However, there are few similarities between the ceremonies allegedly practised by Maasai and Kikuyu. It is almost impossible to date these practices and determine which came first. The Maasai had reportedly sworn brotherhood with the Kikuyu from time to time, in order to open up trade.[45]

The *East African Annual 1948–49* carried a two-page photo-spread showing the Maasai and Kikuyu 'swearing the ancient peace oath'. No date was given, but the ceremony presumably took place that year. It had been arranged by the Kenyan government, and was allegedly mod-elled on some traditional rite – 'on the lines of those which used to take place before the Pax Britannica came to Kenya'. The aim was to stop 'boundary bickerings' between the two ethnic groups, and make 'real peace'. The article described a marvellous concoction and media event, covered by the Kenya Information Office and the BBC, whose commentator was also pictured, 'his microphone being hardly less charged with romantic wizardry for most of the crowd than the incantations of the witch-doctors'. The ceremony was described as unique. It was certainly that – a piece of public relations spin, in which Kikuyu and Maasai were only too happy to collude if it meant a free meat feast and a gathering of the clans.[46]

   The Maasai–British pact may not have been dissimilar to this, except that it was not, by all accounts, orchestrated by government. Its long-term resonance, at least among certain Maasai, undermines White's claim that many such alliances between Europeans and Africans in this period were carried out in an 'inconsequential' way, and did not create new categories of relationship.[47] As for its origins, I suggest that the essential ceremonial ingredients were already there in *en-kiyieu*, the sharing of the brisket. Maasai leaders may have customised a ceremony in a one-off action to suit the new demands of *rapprochement* and survival. Likewise, the white participants may have grafted on to *en-kiyieu* pledges to suit themselves and the new circumstances of the day, in much the same way as Lugard did in order to make friends and influence people who posed a threat to British rule and settlement. It was politically expedient for both parties to invent a tradition to suit themselves, or at least come up with a new hybrid, hence some of my informants' use of the word 'trick' to describe what both parties were doing. The result was a grand pastiche drawn from disparate rituals known to British, Nilo-Hamitic, and Bantu groups in this region – blood-brotherhood, mutual adoption, the sharing of the brisket and peace making.

What can it mean?

Whether or not the blood-brotherhood ceremony actually took place, the centrality of this story in the oral history of the moves and resistance to them is crucially important. Belief in the oath and the honourable behaviour of those Maasai who participated in it may allow present-day Maasai to symbolically regain control of events which spiralled out of their control at the time, and to salvage something good about one element of a bad, bitterly disappointing experience – colonial dispossession. If it is a myth, that is its function. If it is true, it is a hidden and vital beacon in the Maasai–British discourse. Like the 'myth' of Entorror as a pastoral idyll, free of pestilence and problems (see Chapter 5), the British–Maasai brotherhood may represent an immovable truth in a century of turbulence. Belief in it also helps to explain why the Maasai never violently resisted the British – because, say elders, in taking the oath the two parties had agreed not to fight, and to behave honourably towards one another. Although the British broke their side of the pact, by using force to move them from Laikipia, the Maasai reason that this did not warrant a violent response. Ole Gilisho had agreed not to fight the white man again, and the oath therefore overrode the provocation of the second move. As informant Lerionka Ole Ntutu suggested: 'Because of the oath, they could not object to what they were told [to do].' Ole Nchoko, first plaintiff in the Maasai Case, also believed himself bound by its terms, according to informant Ole Ndonyo: 'Ole Nchoko never wanted to fight the British again because they had already shared the blood. So those are the ones who ... because they were unable to do it with spears ... took the case to court because of that anger that they had before.'

It may also partly explain why, having launched the legal action, Ole Gilisho never went to court. Informant Ololoigero said: 'He refused because of the blood oath they had taken with the white people – that is one reason. As an elder would ask himself, "That issue I had already resolved, would I go back to it? How could I go against resolutions which I had made before? I had already resolved those issues with those brothers with whom I had taken the oath." So Ole Gilisho therefore refused to go again to the case against the white people.'

The centrality of meat feasting in the blood-brotherhood ceremony fits a pattern in Maasai rites of passage, particularly for men. Meat feasting with age-set peers marks initiation, ritual emergence from seclusion, ritual removal from and return to the community, induction into warriorhood, and other significant events in the life cycle, including death. All major ceremonies are versions of a sacrificial meat feast called 'passing through the ox' (*empolosata olkiteng*). The ultimate high point for elders is the feast of the Great Ox (*lool-ba*), their final renunciation of warriorhood. Consumption of the upper innards of the animal, particularly from the chest region, is reserved for men. According to Spencer, the chest has a special association with emotional power and fathers, and meat from the centre breast, as well as the tongue, is deemed to be very high status. The brisket fat (*en-kiyieu*) is most important of all: 'there is a reverence about sharing the brisket-fat, as if it is the age-set itself and not just an age mate that is sharing.'[48] This may go some way towards answering the question of whether the ceremony simply bound the participants, or the age-mates of the Maasai and the British as a whole. However, Olonana was not an age-mate of Ole Gilisho's, and neither was Colvile, though Delamere was close (born 1870, while Ole Gilisho was born about 1875). The general consensus among my informants was that it bound the Maasai and the Europeans in their entirety.

Social metaphors

Whatever the nature of any blood-brotherhood ceremony between the two communities, it is not surprising that this should subsequently serve – for the Maasai – as a metaphor for relationships between British and Maasai. In the later colonial period and post-independence, talk of betrayal became rife and remains so today. My informants say that the Maasai believed they had loaned their land to the Europeans and it would be returned at independence; they did not understand why it was allowed to slip into the hands of Kikuyu, Kalenjin and others. Other ingredients of metaphor are embedded in stories of the blood-brotherhood. These include the facts that the Maasai believe they initiated it, in contrast to the official Agreements which were forced upon them. It was verbal and sealed in blood and meat, not written documents which were beyond their control and had no ritual significance. Both parties were men of cattle, therefore admirable. Both parties included age-set spokesmen, or their equivalent. They chose each other in preference to doing a deal with anyone else. It is believed to be unique, and therefore not sullied by repetition. It belongs to an age of 'chivalry' whose like will never be seen again. Also, for those people who believe Olonana took part, it was ritually blessed by his presence, and his collusive role in the second move can be forgotten or at least obscured. Finally, in contrast to the stereotypical image of the warlike Maasai, a willingness to make peace is central to the collective memory of this act. The Maasai have been reinvented here as peace makers, demonstrating a generosity of spirit which the British ultimately abused.

The story of blood-brotherhood was nurtured and reinforced after World War I within the kinds of relationships between certain white settlers and Maasai employees which emerged on settler farms. This idea is developed in the following chapter.

7
Highland Games: Settlers and Their Farm Workers

> Nasoore [Colvile] was a very good man to the Maasai. He was told by Delamere: 'If you want your cattle to increase, employ the Maasai to take care of your livestock.' And the cattle of Nasoore became ten million.
>
> Swahili Musungui, assistant to Gilbert Colvile

This chapter will explore in more depth the real and symbolic nature of relationships between leading settlers and Maasai. Little serious research has been done on the personal interactions between settlers and Africans in Kenya, let alone those specific to the Maasai. There is of course a genre of sentimental settler diaries and memoirs that talk fondly of relationships with domestic servants; they could be called the 'my faithful boy' genre. At the other extreme, scholars have made sweeping remarks about the innate brutality of relationships between employers and workers. My evidence, drawing heavily on Maasai and settler oral testimony, suggests relationships of greater complexity and nuance than have been previously acknowledged. The alleged blood-brotherhood pact may help to explain this. In exploring these relationships, one also finds more evidence of Maasai involvement in the labour market post-1913 than other studies have acknowledged, evidence that substantial numbers returned to the highlands from the Southern Reserve after the second move, and intriguing reports of criminal collusion between settlers and their favourite henchmen against the colonial state.

Lord Delamere, Gilbert Colvile and Parsaloi Ole Gilisho emerge as the odd threesome from the blood-brotherhood narrative. These two settlers were among the first Europeans to dispossess the Maasai from their grazing grounds in the Rift (Delamere settled in 1903, Colvile about 1908), and yet they appear to have forged a close relationship with Ole Gilisho and other Maasai, and are still spoken of today in the warmest terms by many elderly informants. Ole Gilisho's son Mapelu was one exception. Ole Mootian was another, which is understandable given his literacy, heightened political consciousness and experiences at the hands of the British, which included being detained during the Mau Mau emergency. He described Delamere as a snake, 'because he is here and he is there. He was like a big rat. And another one called Colvile, he was pretending to be a friend to the Maasai, but he was also a snake. Let me tell you one thing: these white people who liked the Maasai, it was not the Maasai they liked, it was their land that they wanted.' Certainly, Delamere had had his eye on Laikipia as ideal sheep country from the moment he first arrived in the country from the north in 1897 with his friend Dr Atkinson. His first application for land there was turned down, but in the end he got prime cuts of the Rift and Laikipia, too – the best part of Entorror.[1]

Delamere needs no introduction. Huxley's two-volume biography is the authoritative text, though many would challenge her claim that he was the man who 'made Kenya'. Much less is known about his friend Colvile, the beef baron. Born in 1887, he was the son of Major General Sir Henry Colvile of the Grenadier Guards, who fought in the South African War and took part in several key military campaigns in Egypt and the Sudan. The family had a 3000-acre estate at Lullington, Burton on Trent, and other English properties. Colvile senior was an intelligence officer on the Nile Expedition of 1884–85, which failed to rescue General Gordon at Khartoum. He won promotion and mentions in despatches for his intelligence work in the Sudan, and was decorated for leading the successful campaign against Kabarega, King of Unyoro, writing about these exploits in books that combined travelogue with military history. He was Acting Commissioner of Uganda from 1893–95, but his later actions during the South African War ruined his career. In 1907, just six years after his early retirement to Bagshot, Surrey, he died when his motorbike collided with a car.[2]

Gilbert's mother was Sir Henry's French second wife, born Zélie Isabelle de Preville. According to settler Richard Gethin, who had first met Gilbert in England in 1906, Lady Colvile and her son initially came out to BEA on a shooting safari, to help him recover from his rejection by the Grenadier Guards. He had been offered a commission by the Guards, but then lost some toes in an accident while shooting rabbits, a few days after Gethin met him. (A questionable story, as will be explained.) As Gethin put it: 'Lady Colville [sic] thought it would be a good idea to get Gilbert away from England for a long period as he was taking it badly not being able to join the Guards. She wrote to me in Ireland [to say] she would very much like any information about the country ... as she and her son had arranged to sail in the near future to Cape Town and go on a big game shooting safari through Rhodesia and the Congo to Nairobi.' They liked BEA so much, they decided to settle. By the time Gethin arrived to manage Powys Cobb's farm, Loydien, Colvile was already there, learning about farming. 'I don't quite know who was supposed to be teaching him his job.'[3]

As a widow, Lady Colvile ran a hotel near the railway station at Gilgil that was much frequented by European settlers including Lord Erroll, with whom Gilbert became friends. From his Eton days, he already knew fellow settlers Lord Francis Scott and Sir Jock Delves Broughton, one-time chief suspect in the Erroll murder, and a husband of Diana, later Lady Delamere.[4] However, unlike his class peers, he tended to avoid company and had a horror of the party-going set, preferring to withdraw to his farms and flocks. 'Single-minded in his devotion to cattle and sheep,' writes Trzebinski, 'Gilbert Colvile lived the life of a recluse, eschewing women entirely.' This is not quite true. He married Diana, after she left Broughton and before she wed, with Colvile's blessing, his friend Tom (son of the third Baron) to become Lady Delamere.[5] But his tendencies were monk-like in some respects. '[He] cocked a snook at the establishment and adopted Maasai culture, chewing their snuff and forgoing all creature comforts as their Spartan lifestyle demanded ... By the end of the thirties, Gilbert was such a recluse that he would send his ox-wagon to Naivasha with a servant to pay his bills, fetch his post and bring back supplies to his farm.'[6] Huxley says Colvile 'became a Masai addict', like Delamere and Galbraith Cole, and 'shared their [Maasai] indifference to comfort'.[7]

When one of his former farm managers, Desmond Bristow, first knew Colvile in the 1950s, he owned more than 35,000 cattle on seven farms, making his one of the biggest beef cattle operations in the country (see Chapter 5).[8] Colvile's farms were 'Ntapipi', Kedong, Lariak, Ol-Murani, Sipili, Enkurare and Kongoni. After his apprenticeship with Powys Cobb, Colvile worked for another settler farming family, the Hopcrafts. According to Bristow, Colvile got started on his own account when his mother gave him a farm at Ntapipi (meaning clover in Maa, located west of Lake Naivasha) and £100,000 as start-up capital – a fortune in those days. 'The first thing he did,' said Bristow, 'was to buy out his other three neighbours. Then he started straightening corners out by just excising bits of Maasai land, having come to an agreement with a chief called

Natole.' His farming empire grew from there. He inherited more land from his mother, and 'swapped' it for Lariak, Sipili and Ol-Murani. Colvile was the first chairman of the Kenya Meat Commission, an active member of the Stockbreeders' Association and other settler and farming bodies, but unlike Delamere he did not dabble in politics *per se*. In anecdotal evidence, Bristow and the current Lord Delamere describe Colvile as an eccentric who chose to dress in home-sewn shorts made from Tommy (Thomson's gazelle) skins, lived almost entirely off meat and bananas, laced with the occasional glass of sherry, and could be found of an evening sunk in a high-sided leather chair built like a pub snug, taking Maasai snuff while burning the shrub *o-leleshua*, whose aroma filled the room. He had adenoids, and spoke very strangely when he spoke at all. Unlike John Hurt's slightly more glamorous portrayal of Colvile in the film *White Mischief*, 'he looked like a tortoise with its head out of its shell'.

Colvile was the classic penny-pincher who made pounds grow. Though worth millions, his home at Ntapipi had no modern conveniences. (Huxley said he left more than £2.5m; Delamere, on the other hand, died with debts of more than £230,500.[9]) Bristow said: 'Colvile didn't miss anything when it came to money. He was known to be very mean, in his own lifestyle, the way he paid his ranch managers – it was pretty well much an inherent part of the Delamere–Colvile hierarchy that it was deemed to be a privilege to work for them. They just paid you what they liked, within reason.' His favourite farm and cattle were on Entapipi, where he lived. 'That was his flagship, and he used to commute from there [to his other farms], with Swahili [Musungui] in the back of the car.' He is survived by his adopted daughter Deborah (known to friends as Snoo) who sold all the Kenyan farms, buying a property on the coast and a farm in Hampshire. She was only 18 when Colvile died.

Delamere and Colvile reportedly admired the Maasai, felt some mutual bond with fellow aristocrats, and defended their right to remain bloody-minded outsiders. They, above all, fit Tidrick's analysis of the way in which certain British administrators and settlers admired the supposedly aristocratic qualities of the Maasai – intelligence, bravery, consciousness of superiority, athleticism, good manners, lack of servility and so on.[10] When the government tried to forcibly conscript the warriors, 'Delamere was quick to come to the defence of his beloved Masai'.[11] He acted as mediator between the Maasai and the government after the battle at Ololulunga in 1918. From time to time, the pair also appear to have preferred to 'hang out' with Maasai than Europeans, as stories about shared meat-feasting and, in Colvile's case, alleged cattle raiding, bear out. Deborah Colvile said: 'My father, an only child, was a bit of a loner, not very social, unlike many of his contemporaries. I think he probably felt he had a common interest in cattle, farming, and nature generally with the Maasai.'[12]

Huxley describes how Delamere 'preferred talking to the Masai to the company of most white men'. He would hold nightly sessions by the fire in his living room, to which half a dozen herders were invited. The smell of 'naked limbs smeared with rancid butter' would overwhelm other house guests, but they dare not say a word for fear of Delamere's notorious temper: 'Nothing annoyed him more than to hear his favourite Masai ridiculed or abused.' He even indulged their open raiding of his cattle – a common occurrence, despite the alleged blood-brotherhood. This was a kind of mutual admiration society: 'Members of this tribe were by no means ready to work for any European who asked them to. They were very particular, and only entered the service of men whom they liked ... The Masai took to him from the beginning.' She tells the story of how some young Maasai men approached Delamere at his Njoro farm, shortly after his first sheep had arrived (he imported Ryland, Lincoln, Border Leicester and Romney Marsh rams to put to the native ewes). They said they had heard that a white man had come, bringing many sheep and cattle, and claiming to know more than they did about herding. They were curious to see him, and enquired how long he was staying. ' "I shall stay for ever," he replied. "Then", they said, "we will look after your sheep. You do not understand the pastures. You do not understand sheep. We will help you." '[13]

The story has biblical overtones: Delamere was the chosen one, saved from the wilderness of which they were the keepers. He rewarded their faith in him by spoiling them rotten. Maasai herders were paid more than double what his agricultural workers got, and their food rations were rice and ghee while everyone else got maize meal.[14] Huxley says he had special greatcoats made for herders who came to Njoro, in case they felt the cold. Bristow claims Delamere gave them bowler hats. These descriptions fit Sandford's claim that the few settlers who employed Maasai 'were rather inclined to treat them as pets'. Narok DC E. B. Horne took the same line: 'Steady employment with small owners and business concerns undoubtedly improves the Masai, but residence with certain wealthy farmers of whom they are spoiled pets usually sends them back to the Reserve as reactionary and more swelled-headed than when they left it.'[15] Delamere exclusively employed Maasai as stockmen; to this day, about half the workers on Soysambu are Maasai.

His grandson said of him:

> He found them the only people that he could really understand and get on with. He thought ... that the Maasai were the right sort of people, because if they were going to be tiresome, they would be tiresome to your face. They'd come and steal your cattle and run off with them, and if you shot a few and said 'Don't do that', 'Fair cop,' they said, and that was that. So he had an energetic relationship with them, but he could respect them. And he found them absolutely priceless, as we still do today, as herders.[16]

Ironically, white farmers enabled many Maasai to return to their old stomping ground after the second move, by giving them work and paying them in cattle. In the early days, some farmers allowed employees to graze their cattle alongside their own; the current Baron says this practice was discontinued when the sheer numbers of Maasai cattle meant they could no longer be supported on the same land. (Bristow, however, denies that Colvile and Delamere ever allowed this to happen.) This is partly why some settlers are fondly remembered today. They also spoke the shared language of cattle and sheep. Some, like Delamere, Colvile, and allegedly Galbraith Cole, bothered to learn the Maasai language, too. As Ole Gilisho's son Shoriba put it when asked if his father had had any support from white people: 'The only friends Ole Gilisho had were the white men who took care of cattle.' White pastoralists new to the country soon realised that they could not function without African herders, and the Maasai were superlative in this regard. They also knew these particular pastures, water points, salt licks, climate and diseases specific to the territory and its stock. They could, for example, have warned the Rift farmers about the dangers of Nakuruitis, a mineral deficiency which caused them to avoid certain pastures at Nakuru and Njoro.[17] They later advised Galbraith Cole to move his sheep, who were dying from tick-borne diseases at Kekopey in the Rift, up to Pingwan on Laikipia.[18] So no sooner had the Maasai been expelled than they were asked to return north as labourers, or they simply showed up at the farm gate. A few seasons in Ngatet (western Narok) were enough to convince them that the place was cursed with diseases of both humans and stock. Those who could leave, did so.

Retaking the highlands

From evidence given by farmers to the Native Labour Commission 1912–13 (the NLC, to which Leys and McGregor Ross also contributed), it is clear that some 'northern' Maasai did not join the move south at all, but were already working on white farms by that date. They kept their toehold in the highlands, and were joined by others as the years went by. Some 'Dorobo' did the same, both then and later, when threatened with a forced move south in 1937. Not all the NLC witnesses gave the ethnic background of their workers, but some specifically stated that they employed Maasai. They included, from evidence taken at Nakuru, J. K. Hill, managing a 3000-acre farm at Gilgil for the East Africa Syndicate; H. W. Keeling, a partner of Robert Chamberlain's, farming 20,000 acres at Elmenteita; C. A. Chaplin, a sheep farmer on 16,000 acres at Naivasha; and Delamere, who employed about 100 herders, 'chiefly Masai', on his stock farm.[19] It is not possible to quantify the numbers of Maasai who returned north without further research.[20] But the testimonies of several returnees seem indicative of a wider phenomenon, and I shall go on to examine these.

This may not sound extraordinary. But the so-called recolonisation of the highlands by Africans after 1913 has been described as a largely Kikuyu venture. Studies of squatters and out-migration from reserves have also focused almost entirely on the Kikuyu experience and overlooked that of the Maasai, who seem to have become both squatters and contract labourers on European-owned estates. Bruce Berman writes about the attractions of squatting to 'Africans, primarily Kikuyu, seeking relief from mounting pressures in the reserves ... Squeezed by the chief, squatters found "land and freedom" in the realm of the settler, where the large concessionaires in particular made few if any demands upon them.'[21] For the Maasai, read cattle and relative freedom. African pastoralists are not even mentioned in Berman's précis of the administrative breakdown of Kenya into three labour zones in the late 1920s, in which 'the second zone of Ravine, Nakuru, Naivasha and Laikipia was the heartland of white settlement in the Rift Valley and involved primarily maize and stock production employing Kikuyu squatters'. He repeats the old cliché about Maasai economic conservativism, writing of 'the resistance of nomadic pastoral peoples to change and incorporation in the colonial economy'.[22] Considerable numbers changed and were incorporated, but have become invisible. However, they are in the statistics. For example, 25.28 per cent of adult Maasai males were in paid employment by January 1927.[23]

The main reason why young, unmarried and propertyless Maasai men went out to find wage labour in this period was in order to get the stock they needed to start a family. Stock was their marital bedrock. (Some didn't work for it, but simply stole cattle from European farms and ran it back to the reserve through the forests.) They may also have been coerced by 'chiefs', ordered by government to meet settler demands for labour. Avoidances were a major spur to out-migration, and these included avoidance of recruiters, cattle disease, conscription (the Maasai managed to avoid this altogether), taxation, compulsory labour, chiefly controls, and stock fines under the Stock and Produce Theft Ordinance 1909 (several amendments followed) and the Collective Punishment Ordinance 1909 which exacted collective punishment on villages and sections for cattle rustling and other crimes.

As for taxation, neither tribal authorities (government-appointed chiefs) nor tribal police in the reserves could chase emigrants on to settler land in order to extract taxes from them. There was a split jurisdiction between the country's main police force, which operated only in the urban and European settled areas, and the tribal police, who were answerable to administrative officers in the reserves.[24] Berman suggests that many Africans on European farms managed to evade controls with the connivance of employers, and that emigration undermined chiefly jurisdiction, threatening the state's ability to extract tax and labour. Kanogo describes how the 'White Highlands' were regarded as a haven for Kikuyu wishing to escape conscription into the Carriers Corps: 'Here, settlers protected their employees from conscription for fear of losing their resident labour.'[25] However, Christine Nicholls writes that many Kikuyu were called up, and Maasai were able to fill the labour gaps they left in the highlands.[26] Sixty days forced labour each year on government projects, introduced under the Native Authority Amendment Ordinance 1920, was another disincentive to stay unemployed in the reserve. This was its intention; the law succeeded in increasing the flow of labour to white farms. Finally, it is likely that Maasai in the Southern Reserve learned the news from those who slipped back north that land in the highlands still lay unused many years after their expulsion.[27] Those who returned may have been curious to investigate the possibilities.

There is a contradiction between claims that this out-migration by job seekers was largely voluntary, and Berman's description of settler treatment of African labourers as one of 'chaotic brutality'.[28] There were certainly brutal incidents, involving assaults and murder that went virtually unpunished – and Leys set evidence of these before the NLC, pointing out that settlers were free to fine and flog because 'the ordinary European has more power than a magistrate' and 'the law rarely punishes the European, even in brutal assaults'.[29] (The NLC evidence was his parting shot before leaving the country in disgrace.) Leys devoted most of a chapter of *Kenya* to murders of Africans by Europeans, including the Cole case, although he did not name it.[30] However, Cole was great material for polemicists, but he was not the norm. If all settlers had treated their workers badly, none would have stayed, let alone for four generations in the case of the Delameres. More research needs to be done into rates of Maasai desertion, and the reasons for it. But the following testimony implicitly challenges this monolithic view of the terrible conditions of employment on settler farms, and 'the capricious brutality of the settlers' as a whole.[31]

The returnees

Kitinti Ole Sadera was born in Entorror of a Purko father and a Laikipiak mother, moved to Narok and then to Trans-Mara between 1911 and 1913, and returned north as a junior elder to work for Colvile. He said: 'I came here when I was just a man and married with one wife. I was chased from Ngatet by *ol-tikana* [ECF] which was killing many of my cows. I came back for employment.' Having acquired more cows in Entorror, he took them south but they were also 'finished' by ECF, and he returned north again. This time, he stayed put. At the time of interview he had six wives and 30 children.

> I went back to Entorror with my [first] wife – she was carrying a child – and also an old man living next to me. You had to take a letter from the DC of Narok which said where you were going. Then your name was put down and you were told: 'Take this passport to Rumuruti, or to Nanyuki.' When you reach there, the DC from Rumuruti has to put a stamp and sign it to say you are allowed to stay there. You cannot be refused [permission to return], but what you cannot do is go with cattle, sheep and goats. I sold my cows, and bought more when I settled at Entorror.

Under the Native Passes Regulations 1900, and the Rules to Control the Movement of Masai (issued 24 April 1906), Maasai were allowed out of the reserves with a pass from a Collector or Assistant Collector. But these were time limited, could be refused for no reason, and had to be produced on demand. If one was unable to do so, the punishment was jail for a maximum of three months, and six months maximum with hard labour for repeat offenders. The Resident Native Labourers Ordinance of 1918 sanctioned squatting on European farms, as a means of supplying labour. But it made conditions worse for squatters; the period of compulsory labour became 180 days per year at around two-thirds of the salary of contract labourers, and this rule was applied to all male members of a squatter's family over the age of 16. An amended ordinance followed in 1925, making a squatter's failure to do sufficient work a punishable offence, and prohibiting continued residence on settler farms to persons not under contract. Male migrants were also bound by the much-hated Registration of Natives Ordinance 1915, which required them to carry an identity certificate or *kipande* that included their employment record. Under this system, a worker who left without being signed off by his or her employer could not legally start working for someone else and was regarded as a deserter. Another important piece of legislation adversely affecting labour in this period was the Masters and Servants Ordinance 1906 and its subsequent amendments, which tended to protect employers rather than workers. Breach of the 1916 Ordinance could land an errant worker in jail for up to six months.[32]

Though they did not accompany him on the journey north, other members of Ole Sadera's age-set trod the same path at around the same time. 'Many of the Il-Terito were employed by Nasoore [Colvile], and they bought many cows and they got families through that kind of employment. [In other words, they could acquire brides through cattle bridewealth, and subsequently produce children.] I was not alone – many Maasai boys joined Nasoore's work. We not only worked for Nasoore; we worked for many whites who have settled at Entorror ... They saw that only the Maasai boys can guard these cows, and then we were called.' This 'calling' parallels Kikuyu squatters' descriptions of how Delamere sent for them: 'Lord Delamere, head of the white clan, *mbari ya nyakeru*, and *munene wa mathetera* or chief of the settlers, then invited Kikuyu to colonize these "white highlands". Squatters had not looked for work; he had sent for them, fetching their families and flocks by train ... Pharaoh had sent for Joseph; Delamere called in Kikuyu.'[33] Both settlers brought in Kikuyu to do the agricultural and technical work, leaving the herds to the Maasai. When asked what Colvile was like as an employer, Ole Sadera said:

> Nasoore was a good person to work for. He didn't make noise to workers [i.e., he did not shout at them] because he was a Maasai. He understood Ki-Maasai completely ... He was a good man who was taking care of us, and we were also taking care of his livestock at a place called Lariak.

A few of the diaries kept by Colvile's farm manager at Lariak ranch survive from the 1920s.[34] They confirm patterns of Maasai employment, and other fascinating details of daily life on the farm. A list of labourers for 1928 and 1929 includes such Maasai names as Lataiya, Sotian, Naigisi, Sagamu, Ole Mbweigai, Ole Gerema and Kotel, and the dates of hiring are given. The pay for herders was four shillings a month with *posho*, but 'gifts' of cattle probably made up for the paucity of cash. This is half the average salary cited by Leys for labourers in Kikuyu Province in 1924, without food.[35]

Cattle were regularly dipped as a precaution against ECF. Entries indicate that employees' cattle were also dipped, which was a major plus. Alongside Colvile's stock, workers' cattle also appear to have been inoculated against a range of other diseases including BPP, for which Colvile made his own vaccine. They were also branded, which partially safeguarded them against theft. The entry for 29 October 1928 reads: 'Samburu raiding again, three tried to get away with Masai cattle yesterday.' Amid cryptic reports of locusts, forest fires, weather extremes ('hell and then some?') and cattle killed by lion, farm life appears to have been a daily battle against stock disease.

Colvile's salaries may have been low, but other settlers paid Maasai and other herders a higher rate than their agricultural workers. Besides Delamere, Chaplin, for instance, paid Maasai eight to ten rupees a month while Kikuyu rates started at four.[36] Overall, Maasai herders were either well paid or enjoyed special benefits. These were veterinary protection (when there was precious little in the Southern Reserve), the chance to learn about new veterinary techniques, and to build up a healthy herd of your own while being paid and sheltered to look after someone else's. Informant Ole Njapit, when broadly asked what he knew about Colvile, said immediately: 'He was the one who brought the medicine for cattle diseases. He injected the cattle with the medicine for anthrax and rinderpest, and those medicines were very good – cattle never got sick. Even the dipping, he was the one who introduced it to the Maasai.' No mention of the vets – this was seen as Nasoore's doing. Colvile even visited Ole Gilisho in the Southern Reserve in 1928, when his cattle were badly affected by streptothricosis, took smears and forwarded them to the government laboratory. He appeared to be doing his old friend a personal favour.[37]

Clayton and Savage hint at these benefits, when writing about the rural equivalents of post-war urban attractions and new ways of life – 'the agricultural and veterinary methods, the machines, superior grain and the cattle, poultry and vegetables of the settlers'.[38] But there was another reason why Maasai and other Africans remained on certain farms – the relative freedom from government control that came with being on white-owned land. Berman writes: 'At the district level in the Highlands ... the control of the district officers and commissioners over white landholders was nominal, and their effective authority reduced to a fraction of that which they exercised in the African districts.'[39] In the reserves, DCs and other officials could call on you at any time, to 'see what Johnny was doing, and tell him not to', as Leys put it.[40] But Africans living on remote, white-owned estates had escaped the busybody state to some extent, or at least swapped the DC and his entourage for one master. Of course they were still subject to the law, but they were not under the official boot in quite the same way. In the reserves, closer administration meant abiding by a wide range of new laws. Those young men who fled the reserve may have felt they were escaping tyrannical controls.

Colvile's escapades

When asked about the vaccine Colvile made, Swahili Musungui made an interesting claim. 'Nasoore never made vaccine; he stole it. Even me, I was among the thieves. The vaccine was for rinderpest, and it was for the government vets. Nasoore had so many calves – we maybe took about 200 to the government vet, and the other 300, we used our vaccine which we stole to inject them.' He described in great detail how he stole it, and how his master also stole a book from the vet about how to make vaccines. It is impossible to confirm the truth of the story, though the amount of detail suggests it was not a lie. If so, it indicates a measure of criminal collusion between certain settlers and African employees: they were united in their subversion of government agents.

Other stories about Colvile suggest he went raiding with Maasai warriors into German East Africa during World War I. According to the current Lord Delamere, Colvile avoided joining the army, which was his father's wish, by deliberately blowing off one or more of his toes with a rifle. This contradicts both Gethin's and Huxley's version of events – though Huxley's source, Paddy Grattan, a former employee of Colvile's at Oserian, was more sceptical than she.[41] Colvile had told him the story many times, he said, and he thought it sounded 'rather ... far fetched though amusing'. Delamere is more believable, both because he was close to Colvile, and because Colvile's constant hunting, shooting and sailing indicates that he could not possibly have been severely disabled by this 'accident'. According to Gethin's 1910 diary, he and Colvile went shooting jackal, leopard, impala and other game almost every other day, when they were not busy pig-sticking or sailing on the lake.[42] As for the war story, Delamere confided:

> Gilbert and the army didn't get on. But Gilbert in the First World War realised it was his patriotic duty to do something. He certainly wasn't going to join the army. So he got together with his favourite gang of Narok Maasai and said: 'Now look here, there's a good thing to be made out of this. It's our duty to go and liberate some of the cattle from Tanzania.' 'Whoopee', said the Maasai, 'you're absolutely right. All the cattle in the world belong to us, as you know, but God has unfortunately allowed other people to deprive us.' 'Well', said Gilbert, 'I'll tell you what we'll do. We'll sit up a little *koppie* and see where the Germans are, and when they're at Tuesday we'll do Friday, is that clear?' So they did that, because the Germans are very methodical … you could rely on their patrols exactly … He kept half the cattle, and the Maasai kept half. But because he was a patriot and a nationalist, when he got his half back to Kenya he sold half of them to the British Government who were very grateful because it enabled them to feed the troops. How nice to be so patriotic.

The last line was spoken with heavy sarcasm. He claimed there were several such joint raids, from about 1915 onwards. Again, it is difficult to verify the story; Delamere's closeness to Colvile tends to lend credence.[43] But Grattan told Huxley: 'I got the impression from G. C. that he was involved in World War I Coles (?) Scouts and some sort of intelligence work along the Tanganyika border.'[44] (His questionmark.) There were numerous groups of settler-led 'scouts' involved in war work on the border and inside German East Africa.[45] In his memoirs Gethin described the unorthodox antics of Rosses Scouts [*sic*], led by South African War veteran Major 'Biltong' Ross, who went into German East Africa in about 1914 'with the sole object of playing merry hell … He had enlisted about 40 of the hardest characters in the colony.' The Ross family threatened to sue in 1955 when Gethin's memoirs were serialised in the local press. Gethin had described Ross as 'ruthless', said his scouts were known for 'looting and killing all before them', and that they had been disbanded because 'they were too tough for East Africa even in those days'. This suggests that the line between freelance and authorised raiding was a thin one, and that some settler-soldiers trod a very fine line indeed.[46]

  The Colvile raiding story was corroborated by his closest Maasai associate, Swahili Musungui, without the question being asked. 'Nasoore even led Maasai warriors to raid a community called Il-Tatua. During that time there was also fighting between the British and the white people from Tanzania. During that raid, they went to a place called Magadi and

Nasoore killed two people. He brought a gun from there, and a horse who was given the nickname Sokis [Socks], and the gun they brought was called 303. Even me, I have used that gun.' Again, this seems too much detail for a fib. Further corroboration came from another inform-ant, John Ole Karia, who as a child was, in his own words, 'Colvile's kitchen *toto*' [child]. He said that the alleged raiding all those years ago connects directly to a modern land claim. The Enaiborr-ajijik Group Ranch is contesting ownership of several 'land parcels' near Entapipi, all formerly owned by Colvile. It is alleged that Land Parcel 413 was given to Colvile by the British colonial government 'as a token of appreciation of his 1st World War exploits' – a euphemism for his freelance raiding.[47] How do they know? Ole Karia said: 'Because Colvile boasted about his exploits to the old men. He would say how he had defeated the Germans, how he had acquired the cows and the land. In fact, he shared the loot with the Maasai.' Colvile could not resist bragging like a regular warrior. The only problem was, the land he was given had been excised from the neighbouring Maasai Reserve.

Sandford and Huxley described Lord Delamere's involvement in World War I, when – at his own request – he was placed in charge of several hun-dred Maasai warriors who were employed as scouts and messengers by the Intelligence Department to patrol the German East Africa border.[48] In early 1915, about 100 rifles were given out to warriors of four sections, including the Purko, and these gave them 'more confidence' to scout. In June 1916, they were supposed to hand their weapons in, but the last of them were not handed over until December that year. Sandford noted: 'There is no doubt that the possession by the Masai of these few rifles helped to instigate the majority of the raids which the Masai carried out against tribes in German East Africa.' He listed eight: six in 1915 and 1916, while two were undated. Unsurprisingly, there is no mention of any European sponsoring or joining such raids from the British side. Neither is there strong condemnation. The British had worried about whether they could rely on Maasai loyalty during the war, given their proximity to the Germans. One is tempted to read faint praise of freelance raiding into the statement: 'Not only was there no suspicion of disloyalty on their part, but they also rendered material and voluntary assistance to our forces, and on many occasions proved extremely useful.'[49]

The experience of other employees

Nkotumi Ole Kino was a herdsman for Colvile. Again, his family had been moved south, but he managed to return north to work and has stayed on Laikipia ever since. At the time of the second move, he said, he was old enough to be looking after sheep, which indicates that he was at least 7 or 8 years old. In Narok District the family settled at Orkinyei, near Lemek, but he returned north after *eunoto*, primarily because 'this place is suitable for livestock and human beings, and I know that place is poisonous, especially *enkipai* [Trans-Mara].' Ole Kino decided to play on the fact that some so-called Dorobo had remained in Entorror when the Maasai left, and that British officials could not tell the difference between them.[50] He pretended to be 'Dorobo' in order to be allowed to stay: '[To officials] I never said I was Purko; I just said I was Dorobo, with a family living at Entorror ... I had to lie that I was Dorobo, and then I got the passport to go [north] ... Many people did this.'

This pretence appears to have been a regular practice, and one with a long history. Merker mentioned Maasai warriors in German East Africa fooling coastal caravan leaders by pretending to be poor and 'calling themselves Wandorobo'; they befriended and lulled them into a false sense of security before calling for reinforcements who robbed and killed the porters.[51] Waller mentions how the difference between 'true' and 'temporary Dorobo', as the Nanyuki farmers called them, was recognised by other early writers including Smith and Hobley. Spencer attempted to unravel who was who among Dorobo groups in northern Kenya. More recently, Lee Cronk has discussed Dorobo identity and perceptions of it.[52] The Dorobo of Laikipia are a story in themselves, and one can track their continued sojourn there through the annual reports; for example, there were 150 people classified as Dorobo in the district in 1926. The fact that they were said to own 1000 cattle rather gives the game away; if they were 'true' Dorobo they would not have had any.[53] But Maasai 'become' Dorobo when they lose their cattle, and vice versa. Goldfinch said of those Dorobo who moved south with the Maasai: 'in about a year's time they had all crept back again minus their property'.[54] By all accounts, there was an awful lot of creeping going on. The closure of the government station at Rumuruti during World War I provided a convenient loophole; there was no government presence there until after 1918.

Old friends Salonik Ole Pere and Kireko Ole Kimiri also worked for Delamere and Colvile, their families having returned from the south. At the time of interview, they still lived in Rumuruti. Their stories repeated the refrain of mutual cooperation and benefit flowing from employment by settlers, which contrast with what are perceived to be the greater tensions and deprivations of the present time. 'These people

[Delamere and Colvile] were together with the Maasai and they were helping each other how to rear cattle,' said Ole Kimiri. Ole Pere added: 'The white people ... really helped us, especially the poor, because they were giving food to the poor, providing employment in their places, and much other help.'

To return to the late Swahili Musungui, it is worth describing in more detail his relationship with Colvile and employment on his farms. Of all Colvile's Maa-speaking employees, Swahili was probably closest to him, having worked with him from boyhood up until Colvile's death in 1966 at the age of 78. Huxley, from information supplied by Paddy Grattan, described how Colvile turned to Swahili towards the end of his life, when he was on his own again after Diana's departure – into the arms of Tom Delamere. 'He reverted to one former custom: almost every evening his Dorobo headman, oddly named Swahili, would squat down on the living room carpet and converse for an hour or so with his employer in Maasai.'[55] Swahili himself told me proudly: 'I became almost like a son of Nasoore. He could not go a step without me. I was the interpreter, the one who interpreted things to the Maasai. For example, every judge has his own interpreter. So I was the interpreter of Nasoore.' In those days, 'Delamere was like Moi now, and Nesore was like Saitoti now [the then vice president of Kenya]. The two of them were people of one house.'

Swahili's family was not directly affected by the second move, having lived in the Mau area for generations. 'I was born at Entapipi. When I grew up [but was still an uncircumcised boy] Nasoore employed me. The first work that I did was feeding dogs, the ones which were used to kill lions, and my salary was three shillings a month.[56] When I became a grown-up, the salary was raised to 15 shillings.' His second type of work for Colvile involved 'shooting buffaloes so that they didn't bring diseases to cows'. By this, he would have meant primarily ECF. He remembered going on a fortnight's training course in Naivasha to learn how to shoot. 'I was shooting the wild animals to feed the dogs. When the lions killed livestock, I was the one to shoot.' Swahili was circumcised in 1938 (he clearly remembers war breaking out the following year), and says he did this second type of work for the next seven years. Later, as interpreter and gunman, he accompanied his employer when he drove from farm to farm.

Workers were paid in cows: 'They wrote down the days of your work. When it reached the fifth day, heifers were put aside for the Maasai. You were told to get in and choose the one you wanted.' Swahili was asked why he had liked Nasoore so much, when he was one of the people who had taken over large areas of Maasailand:

> He was a very, very clever man. He knew very much how to cheat the
> Maasai. He was helping very much the people who have been work-
> ing on his farm, because those people who came when they were
> poor, they went with so much property [cattle]. So he assisted many
> Maasai, because you came when you were not married, you got a job,
> you got money, you could now afford to take care of that wife and
> children. So we have liked him very much because of that. The cattle
> you bought, you put them on his farm. When you wanted to leave,
> you went out with 200 cows, 800 cows, and when you had joined
> Nasoore's farm you had no cow. He was ... helping the Maasai while
> slowly taking their land.

Swahili also believed that when the first Lord Delamere died, he ordered
Colvile to 'take care of the Maasai'. He credited Colvile with having prevented settlers from taking any more land from them: 'It was Nasoore
who stopped the white men from settling in Narok.' In Swahili's eyes,
Colvile was effectively guardian of the boundary. Unbeknown to him,
but according to Bristow and the current Lord Delamere, the settler had
also fixed several boundaries to suit himself.

When speaking generally of his attitude to whites, Swahili said
something that echoed Delamere's earlier remarks about his grandfather's respect for Maasai directness, even when it amounted to brazen
cheek and criminality. 'There is one thing with the white people. If
they tell you, "I like you", they have to show that they like you. If they
tell you they will beat you, they beat you there, there, there, on the
spot.' He appeared to be saying: 'At least you know where you are with
them,' and his tone was quite approving. At the time of interview, he no
longer seemed to know where he stood socially, hence his nostalgia for
the old scheme of things. Under Colvile, he had enjoyed some status in
the settler's farming empire. He mediated, his translations carried
weight, he travelled around in a large car, handled big guns, and most
importantly was seen to be a friend of Colvile's and party to his secrets –
including his relationships with women. Swahili had plenty to say,
for example, about his master's triangular relationship with Diana and
Tom Cholmondeley, and aired his theory about the Erroll murder, inti-
mating that he had inside knowledge that a hired Somali gunman
had done it. (The current Lord Delamere claims that the killer was in fact
his stepmother, Diana. In interview, he cheerfully declared: 'She was a
shootist!')

But all along, if we are to believe the whispers, he was 'just a Dorobo bastard'. Now he had reverted to that status. Worst of all in Maasai eyes, he had no children or cattle. In his last few years, he was not a happy man. Colvile had given him a plot of land, where he still lived: 'We became close friends, and I was given it as a gift.' But he claimed that most of it had been taken off him by the government when Entapipi was subdivided for agricultural development. The final insult was that he had been fenced in by a neighbouring European farmer, in an apparent attempt to force him altogether from contested land. He begged me to ask Bristow to give him a job on Solio, and dictated a letter for me to pass on. Within a year he was dead.

To conclude, oral testimony about an alleged blood-brotherhood has taken my inquiry well beyond the moves and court case. Yet the relationships revealed are important in understanding Maasai strategies and responses to colonial rule, as well as their 1913 legal challenge. The Maasai returnees who reversed the exodus from Entorror may have had the last laugh, except of course that they were left without land titles at independence and afterwards. They and other Africans exploited one of the fundamental contradictions of the colonial state in this period – its desire to separate Africans from whites by creating reserves, while at the same time meeting settler needs for African labour, which meant allowing them out again and encouraging intermingling.

The overwhelmingly warm endorsements of Delamere and Colvile quoted here cannot be squared easily with the assessments of people like George Goldfinch, or Ole Mootian. The bald fact is that they and their ilk helped to dispossess the Purko and other victims of forced moves, as Goldfinch explained when warning the Anti-Slavery Society to view with suspicion any Maasai Commission that had Delamere and Berkeley Cole as members:

They both I think like them [Maasai] as individuals and employ a lot as herd boys but they are absolutely unscrupulous as regards natives and would sacrifice any tribe for the sake of getting hold of a few thousand acres of their land. In fact they are really at the bottom of all the Masai troubles and but for them the Purko would never have been shifted from Laikipia, and the recent disaster to the Uasin Gishu lot is simply due to Lord Delamere.[57]

For an explanation of the effusive Maasai praise of particular white men one must look to what blood-brotherhood represented, and consider the advantages of working for certain settlers. It goes without saying that these relationships were unequal and paternalistic. But it is not good enough to dismiss them as unremittingly exploitative and brutal. Exceptional characters such as Swahili Musungui saw more of their masters than their wives did, and gained many benefits. His relationship with Colvile appears to have been close and collusive. Though they may have deluded themselves, both parties appear to have believed that they cared for each other. Alternatively, the Maasai–settler relationship may be interpreted as one of symbiotic alliance rather than dependency, with features that mirror the earliest Maasai–British alliance and (apart from caravan guides and interpreters) the earliest form of Maasai contract labour – as hired warrior auxiliaries on punitive raids. They, too, were attracted to this work because it paid cows not cash, and gave them a measure of freedom from the control of elders. Though cash came later, it was clearly not the main reason why Maasai went out to work.

There was more to Delamere and his cronies than either Huxley's airbrushed biography, other scholarly studies of the period, or critics of settler society such as Leys and McGregor Ross have suggested. To go beyond essentialism, maybe they were neither heroes nor villains but something much more rounded and interesting, as 'the interpreter of Nasoore' suggested. As for the 'myth' of the blood oath, we shall probably never know the truth. The secret, if there was one, was buried with the odd threesome.

Conclusion

> The time has come for the Maasai community to pause and review, reflect and evaluate their total losses during the horrific removal by the British Imperial Regime from their lands ... It would not to be improper or imprudent for the Maasai to demand that they get back part of their pasture lands or be paid compensation ... .[1]

In evicting the Maasai from the Rift Valley and Laikipia, the British clearly perpetrated a great injustice that has repercussions to this day.[2] After all, Britain made a solemn contract over land and broke its terms just seven years later, under the pretext that the Maasai themselves had asked to be relocated. The numbers of people who died during the first phase of the second move (up to August 1911) cannot be proved and may be negligible; the injustice goes deeper and wider than that. The northern sections lost the greater part of their land, and the wide range of habitat necessary for transhumant pastoralism. The extended Southern Reserve was an inferior substitute for the northern territory. Its western extension lacked sufficient permanent water sources, accessible forests and drought refuges, while disease vectors were more prevalent.

From their own surveys, the British had a good idea of the quality of this reserve before they sent the 'northern' Maasai there, but it was not until Aneurin Lewis's studies of ticks and tsetse fly in the 1930s that thorough scientific investigations were made and a clear picture emerged of the environmental challenges to which certain Maasai communities were exposed. Any subsequent 'overgrazing' and 'overstocking' were a direct result of increased confinement, overcrowding in certain areas, curtailment of seasonal migration, an almost continuous state of quarantine, and early restrictions on cattle trading, although another contributory factor was improved veterinary services, when they eventually came, which led to larger stockholdings.

Despite enormous losses to disease, overall stock numbers rose, leading to concerns about congestion in the reserve by the 1930s. There is clearly a mismatch here between the reality of herd growth, which signifies health and prosperity, and what is collectively remembered – the demise of Maasai herds and society following the moves. This suggests that stories about disease are partly a social metaphor, representing social fragmentation and Maasai loss of control over their physical environment, which were major end-results of the moves and colonial intervention. Another anomaly concerns small stock: complaints about the unsuitability of the Southern Reserve for cattle overshadows the fact that large areas offered excellent pasture and other favourable conditions for sheep and goats, particularly on Loita.

The evidence suggests that those responsible for this injustice were principally local administrators, who privileged the interests of white settlers over those of Africans and Indians. It was not primarily driven by the Foreign or Colonial Offices in London whose officials tended (at least in the early days) to underline the need to respect 'native' land rights and balance these against European land claims. The Foreign Office and the Colonial Office were to a large extent kept in the dark by their commissioners and governors in British East Africa. They were lied to on many occasions over such crucial issues as alleged Maasai acquiescence to both moves, the reason for the second move, and whether or not promises of land in the highlands had been made to European settlers before plans for the second move were hatched. The failure to send Arthur Collyer's 1910 'Report on the Masai Question' to London amounted to criminal deceit; it was held back for some 18 months, rendering its recommendations useless. Poor communications by sea between East Africa and London aided the BEA administration in its deception of the home government, particularly under Girouard. However, local officials were by no means united in their sentiments and did not speak with one voice. Dissenters, including high- and middle-ranking officials who objected to the manner in which the Maasai were dispossessed, though not necessarily to the moves themselves, were overruled and sidelined.

As for the Maasai Agreements, it is doubtful whether the Maasai signatories to the first Agreement understood its implications, but no evidence exists to show that it was actually forced upon them. The second Agreement was certainly forced upon largely unwilling signatories, some of whom, notably Ole Gilisho, had received threats from British officials which they took very seriously. The age-set spokesman clearly saw the implications of the second Agreement for the first, likening the latter to 'a broken weapon which is finished with'.[3] All the available evidence belies Lewis Harcourt's confident claim to the House of Commons in July 1911 when Secretary of State for the Colonies: 'The Masai came to a unanimous and even enthusiastic decision to move to the Southern Reserve.'[4] After the 1913 court case, threats to his life reportedly forced Ole Gilisho to cancel his plans to visit Britain to pursue the legal action before the Privy Council – one of many claims made in oral testimony which do not appear anywhere in the written literature. The second Agreement amounted to an abrogation of the first, since the government broke its promise to leave the Maasai undisturbed in Laikipia 'for so long as the Masai as a race shall exist'. If it was intended to amend the first, it should have been signed by the same people, and not by a minor, 13-year-old Seggi. But by that time, most conveniently for the British, his father Olonana was dead.

Although Olonana's duplicity was a factor in Girouard's plans to move the Maasai for a second time, linked to an internal struggle for control within the Purko section, this was not the overriding issue. Diana Wylie has asked: 'Was the move prompted exclusively by European financial interests, as Leys suggested?'[5] On my evidence, the answer is: 'No, not entirely'. Lord Delamere and other leading settlers certainly lobbied hard for land on Laikipia. But other factors behind the move included a desire to reverse a 1904 two-reserve solution to the 'Maasai problem' which clearly had not worked; to corral, control and tax the Maasai more effectively in one area, and through taxation to produce more wage labourers; to prevent them from wandering between the two reserves; to stop the spread of 'native' stock disease to European farms; and to acquire an area for white pastoralists that was reportedly free of ECF. The Governor allegedly wished to support 'traditional' leadership in the form of Olonana and his sons, who already lived in the Southern Reserve, conveniently close to the centre of government, and who favoured a one-reserve policy because their authority was being challenged as a result of the physical division. Girouard believed that government control of the Maasai would be facilitated by this 'special relationship' to Olonana's dynasty; in fact, the relationship fell apart. Finally, there was some official fear of settler aggression towards the Maasai (and their likely retaliation) if the two communities were not fully separated and desirable land not freed up for whites.

There are many legal aspects of the treaties and forced moves, first aired on appeal in 1913, which would not stand up to scrutiny in court today. These will be tested if the Maasai realise plans to bring another legal action – something veteran Maasai politician John Keen threatened in 1962 when he brought to the constitutional, pre-independence talks at Lancaster House, London, his memorandum on the treaties.[6] In 1913, the British government had good reason to believe it would not win the Maasai Case. Private discussions within the CO, as revealed in PRO archives, show that recourse to the Privy Council was considered well before the case even reached court, and the likelihood of having to pay compensation was also anticipated. Furthermore, close examination of the 1911 Agreement, and connections between it and the granting of a concession to the Magadi Soda Company, suggests that the Company and its successors at Lake Magadi, Kajiado District, probably have no legal right to be there.[7] In the early 1960s, with independence imminent, the British admitted they had a 'moral obligation' to the Maasai as a result of the treaties, but defined this very narrowly. These words are likely to return to haunt them, while PRO files from 1962 on the future of the Agreements, marked 'Closed: no further action to be taken', may suddenly prove to be of great official interest.

Oral testimonies, albeit problematical, augment and enrich what is already known about these events. By adding Maasai voices to the story it is enriched and amplified, and an imbalance in the literature at least partially redressed. It is acknowledged that interviews with members of other sections would produce a much fuller picture, and add a necessary corrective to predominantly Purko claims, but this is beyond the scope of this particular study.

Repercussions

The repercussions of these events in the short term included distrust and alienation; a Maasai retreat from the colonial state which led to their falling behind in development and education (a situation that still persists in many parts of Maasailand, though this is partly attributable to reactionary attitudes, especially to the education and other rights of girls and women); and considerable stock and human losses in the decade following the second move when resistance had not been developed to diseases such as ECF and malaria, which were either unknown or not prevalent in the north. Warrior uprisings in 1918, 1922 and 1935 can also be subsumed under repercussions. The significance of the fatal clash between British forces and *il-murran* at Ololulunga, western Narok, in 1918 has been misinterpreted by historians who have relied upon a single explanation – that it was sparked by opposition to forced schooling.

Although this was a contributory factor, the uprising – which left between 16 and 51 dead, depending on whose account one accepts – seems to have been symptomatic of a wider frustration with British colonial rule and land alienation.[8]

The evidence suggests that Maasai herds on Laikipia were resistant to ECF, and enzootic stability could be maintained. Also, when they had the space to roam, herders coped with disease by simply moving stock away from infected pastures. In the longer term, a combination of factors, driven by the moves and colonial intervention as a whole, led to acute population pressure, land degradation, erosion of subsistence livelihoods, a decline in the quality of stock since quarantine prevented the import of bulls, loss of markets, and increased vulnerability to drought. Furthermore, by placing the Maasai in reserves, and hedging these about with treaty promises that included a pledge to repel ethnic 'aliens', the British nurtured an enduring obsession with boundaries, promised land and exclusivity. This has manifested itself since the early 1990s in 'ethnic' clashes and renewed calls for *majimbo* (regionalism), while a fluid ethnic identity has become increasingly concretised.

Further research would be required to find out what impact the immigrants and their stock had on Maasai communities already living in the Southern Reserve, some of whom were violently displaced. More broadly, to view the moves as something which massively inconvenienced the Purko, but left everyone else virtually untouched, is to overlook the ripple effects on other Maasai sections and neighbouring ethnic groups. In 1904 a stone was thrown into the pond that was Maasai life in the northern Rift, making waves for everyone around.

These negative repercussions must of course be weighed against the gains, both from colonial contact and as a result of moving to the south. Individuals gained from the first of these, in terms of power, finance, education, patronage and wider opportunities. Maasai were forced to make new and profitable alliances, both with agents of the state and other ethnic groups, and between Maasai sections. The gains accrued from moving to the south have included enormous revenues from wildlife tourism, and for some individuals, from lucrative wheat farming and land speculation. However, these riches have not been equitably shared beyond the county councils which manage the game parks in Maasai territory and powerful individuals and families, though some tourism profits are distributed to the community via Maasai-run wildlife associations. Current grievances in Maasailand include widespread corruption, politicisation, monetisation, the widening gap between rich and poor, encroachment by other ethnic groups, expansion of cultivated areas (though Maasai are increasingly turning to cultivation themselves), and the privatisation and subdivision of land under a land adjudication programme into uselessly small plots. In this process, poor Maasai are being dispossessed by rich Maasai and others.

Resistance and power

The form of passive resistance led by Ole Gilisho was extraordinary for its time and place. But it is entirely consistent with the role of an age-set spokesman in leading, counselling and defending the community, sometimes – if necessary – in defiance of elders. In seeking legal redress through discourse with the British, Ole Gilisho was also conforming to role type, and the form of dispute settlement which he hoped to use had parallels in Maasai customary justice. He emerges from this evidence as an unsung folk hero whose personal charisma and leadership qualities, combined with the opportunities handed to his age-set by nineteenth-century battle victories, and later by the opportunities offered by British colonialism to age-set spokesmen and prophets in particular, allowed him to gain unprecedented power and moral authority. He exercised this power to gain ascendancy over the prophet Olonana and his sons Seggi and Kimuruai. Ole Gilisho's alleged 'conservativism' – a term applied to him by administrators, and unquestioningly reiterated by some scholars – can be reinterpreted as progressivism. Though he later appeared to collude with the state himself, as a paid 'chief', he played a double game – promoting Maasai interests while supporting the state in, for example, its attempts to dismantle warriorhood.

Certain anthropological models of authority in the 'classic' pastoralist gerontocracy, which historians also tended to accept pre-Hodgson and others, are implicitly challenged and undermined by this evidence.[9] The model centres on the idea that power resides with male elders and councils of elders; that warriors have no real 'political' power or authority; that warrior rebellion against elders is merely ritualistic in nature; and that prophets are more powerful than age-set spokesmen. The behaviour of Ole Gilisho and his supporters from *c*.1912–13 may be dismissed as an aberration and departure from the norm, but their rebellion was for real, not ritual. The reasoned nature of this resistance belies the stereotypical image of 'wild', irrational and predominantly volatile warrior behaviour. Though produced initially by nineteenth-century explorers and missionaries, this image is still current in some quarters.

As for Norman Leys, he did not manage to prevent the second move. But he helped to bring down Girouard, in sowing the seeds of doubt about him at the CO and supplying information that proved the Governor was a liar. His actions prompted the CO to stop the illegal 1910 move, and he both inspired and fuelled actions by an influential circle of humanitarians in Britain to challenge and discredit colonial policy in East Africa. He did more to assist the Maasai in their legal action than he ever admitted in print, while maintaining a curious ambivalence towards them which was characteristic of the man.[10] Leys also helped to widen the debate about imperialism, the human rights of imperial subjects, land and labour, health and housing, the impact of capitalism upon the developing world, the links between macroeconomic policies and grassroots injustice, and the nature of development in Africa. This debate was a precursor to nationalist struggle and the eventual dismantling of the colonial state – which is ironic, given that Leys did not oppose imperialism *per se*; he simply argued for 'humane' and transparent imperialism on liberal lines. Together with McGregor Ross, he challenged settler hegemony in Kenya and helped to scupper settler attempts to create a white dominion. As a Christian Socialist he questioned the role of missionaries, and demanded quality, dogma-free education for Africans since 'they have as much right as we to understand the world we both live in, and far greater need of knowledge as a defence against oppression', foreseeing that this would 'make them think politically'.[11] Leys warrants a full-scale study in his own right, and I have not attempted to cover the story of his post-Kenyan career. As Cell notes, in semi-retirement 'his home and surgery at Brailsford near Derby became the nerve centre of an intense, unceasing publicity campaign'.[12] Unfortunately, he never lived to see the fruits of it, but African nationalists, the international human rights movement, development practitioners, and a succession of whistle-blowers, have much to thank him for. He had convictions, and was not afraid to voice them.

There was a spectrum of critical dissidence over the Maasai moves in both Britain and BEA, and one must not allow individual contributions to the whole to be obscured by noisier voices such as that of Leys. The dissidents included, of course, Ole Gilisho and his supporters. On the European side, one of the more surprising champions of Maasai rights turns out to have been George Goldfinch – game warden, settler and Master of Foxhounds.

Blood-brothers and reversed exodus

The reason why the Maasai failed to violently resist the second move may partly be ascribed to the blood-brotherhood pact allegedly made between leading white settlers and Maasai representatives sometime before 1911. The oral evidence for this is overwhelming; I believe it did take place, very likely at Soysambu in the Rift Valley. On this point, however, I shall have my cake and eat it too: whether or not the ceremony occurred, the centrality of this story in oral testimonies which reflect collective memories of the moves is crucially important. If a myth, it functions as a very powerful social metaphor; but if it is true, it is a vital and hitherto unseen beacon in the Maasai–British colonial discourse. Belief in it also helped to shape relationships between certain white settlers and Maasai workers on European farms in the highlands after World War I, relationships rich in irony. These were more complex than many scholars have acknowledged, and cannot be easily dismissed as paternalistic, or characterised solely by white brutality. The indulgence of Maasai workers by Delamere and Colvile in particular was not simply a matter of 'spoiling their pets'; this was a two-way street whereon a mutual admiration society formed, embedded in the idea of blood-brotherhood between two peoples, and shared notions of racial superiority. Of course, both these men were also motivated by land greed.

Certain leading settlers enabled, and actively encouraged, considerable numbers of Maasai to return north after the second move and reconsolidate their herds and families on white-owned farms. In this way, many 'northern' Maasai quietly reversed the forced exodus from Laikipia and the Rift. The numbers are unquantifiable without more research. Some retained their toehold in the north all along, either by pretending to be 'Dorobo' or by refusing to move at all, successfully seeking refuge as workers on European farms. This pattern of return and reoccupation of the northern territories refutes the standard wisdom that the so-called recolonisation of the 'White Highlands' was a largely Kikuyu venture.

The legal situation today

In the intervening years Maasai leaders have repeatedly complained about the land alienation and its consequences (see article).[13] My opening quote summarises the view held by Maasai activists who now intend to revisit the treaties and 1913 case in order to seek legal redress and reparations. They are inspired by legal precedents which include the successful challenge of colonial treaties by indigenous peoples, and their reclamation of natural resources. Such treaties were not all bad news for indigenes; some can be used today to prove separate nationhood and to hold former colonial powers accountable for past pledges. (Aboriginal peoples in Australia, who never made a treaty with the colonial state, have long pressured the government to draw up a treaty precisely because it would guarantee their rights and recognise their cultural distinctiveness.)

One of the best-known challenges has been brought by Maori in response to the Treaty of Waitangi, signed by the Crown and Maori leaders in 1840. Unbeknown to them, it took away their sovereignty. But it expressly guaranteed them 'full exclusive and undisturbed possession of their lands and estates, forests, fisheries and other properties'. Through a tribunal, established in 1975 as a permanent commission of inquiry into claims relating to the treaty, Maori are able to lodge claims to land and natural resources, or hold the government accountable for breaches of treaty principles. The tribunal does not settle claims, but can recommend whether or not actions or omissions of the Crown breached these principles. However, in the Kenyan case no such domestic legal process exists, therefore this avenue is not open to the Maasai.

There are parallels between the two scenarios, including the fact that versions of the Treaty of Waitangi were oral. James Belich, who maintains there were at least five treaties, writes that the fourth version was 'a series of oral agreements *among* chiefs, as well as between them and those speaking for the Governor … The trouble is, how do we now know what was in them?'[14] This echoes the oral agreements made at the time of the alleged blood-brotherhood. And as with the Maasai, the illiterate Maori signatories' understanding of what the Treaty of Waitangi actually meant was very different from that of the British. Vincent O'Malley writes of this and other early agreements in New Zealand: 'Indeed, more often it would appear that Maori interpreted early agreements as confirming rather than extinguishing their rights, albeit in a modified environment in which their land and resources would now be *shared with their new guests*.'[15] (My italics.) According to my informants, this 'sharing' is exactly what the Maasai believed would happen as a result of their Agreements. Hence the persistent lament that alienated land did not revert to the Maasai community when the British left Kenya.

Patrick McAuslan, now professor of law at Birkbeck College, University of London, says of the 1913 Maasai judgement: 'My view of the case is the same today as it was when Yash Ghai and I wrote our book in the late 1960s: it is hypocritical and political.' However, he foresees several problems with the forthcoming Maasai action, not least because key precedents – such as the Mabo judgement in Australia, and Waitangi claims – have taken place in national courts. It is dangerous, he says, to assume that British courts will virtually rewrite the law on indigenous land rights just because other jurisdictions have done so:

> First, Australian and New Zealand courts are national courts dealing with national land issues. The British courts would be asked to deal with Kenyan land issues. Since the nineteenth century, it has been established that British courts will not pass judgement on land cases in foreign countries where they cannot enforce it. How could they enforce a judgement that Kenyan landowners and the Kenyan government should restore the Maasai to their ancestral land?[16]

McAuslan also believes it highly unlikely that Kenyan courts will go down the road towards land reparations as courts in other former colonies have done. 'The courts of these countries [apart from South Africa] are staffed by the descendants of the dispossessors and are accepting vicarious responsibility for the defaults of their forebears. But what colonial evils have the Kenyan judges and their forebears been guilty of? They too were dispossessed of land and treated badly.' Lawyers for the Maasai have said they would prefer to seek justice in London, at least initially, but too much time may have passed. The Privy Council is no longer an option; Kenya abolished appeals to this body in 1963. Other possible avenues include the African Commission on Human and Peoples' Rights in Banjul, Gambia, or the use of UN protocols. The UN Draft Declaration on the Rights of Indigenous Peoples states:

> Indigenous peoples have the right to the restitution of the lands, territories and resources which they had traditionally owned or otherwise occupied or used, and which have been confiscated, occupied, used or damaged without their free and informed consent. Where this is not possible, they have the right to just and fair compensation.[17]

Article 14 of the International Labour Organisation (ILO) Convention 169 says something similar about the need to establish land claims processes.[18] But the Kenyan government has not signed ILO 169, and the UN Declaration remains a draft. One major sticking point in discussions of the draft is the refusal of some governments, including that of Britain, to accept the use of the plural 'peoples' in its wording. Britain refuses to recognise the collectivity of human rights, explaining: 'We believe that if states with indigenous communities ratify and implement the six most important UN human rights treaties, they can do more to improve the human rights of indigenous people than by creating new collective rights.'[19] Then there is the whole question of who is indigenous in Africa; some argue that all Africans are indigenous, therefore the Maasai (in this instance) cannot demand special treatment. If discussions are successfully concluded on a new draft constitution for Kenya, this may allow redress for historical injustices including land grabbing. But as this book went to press, the result of a November 2005 referendum on the constitution was not known. Two government-appointed land commissions have reported in recent years, but these review bodies cannot resolve this kind of land claim, any more than the KLC did in the 1930s. It is not within their remit.

At the time of writing, the Task Force of Maa-speaking Communities behind the forthcoming action had submitted a Memorandum on the Anglo-Maasai 'Agreements' to the British government, setting out their historical grievances. No official reply had been received. But Lord Avebury MP, deciding to take an interest in the Maasai as Edmund Harvey and others had done nearly 100 years earlier, pushed the Foreign Office for a written response, after reminding it that there was a case to answer. He was told: 'The legal position today is quite clear: at the time of independence the Government of Kenya inherited any obligations that formerly rested on us as the sovereign power.'[20] There is a distinct sense of déjà vu.

*   *   *

Despite some anomalies, there appears to have been a solid basis for Maasai belief in the intrinsic healthiness of their former grazing grounds. White settlers flocked to the highlands principally because they were seen to be healthy, and offered a welcome respite from the sickness that stalked the coast. It follows that what was healthy for whites was healthy for black Africans, too. But Maasai attachment to Entorror, their former northern territory, represents a larger nostalgia for the past, and in particular for Purko well-being and hegemony over other sections following their rout of the Laikipiak in the nineteenth century. The Purko's last foothold in Entorror, Laikipia, has taken on the status of a lost Eden in social memory. It is said to have been sweet, disease-free, blessed by good pastures and plentiful rain, in contrast to the 'bitterness' of the south. Intertwined with this idea is nostalgia for the concept of a Maasai nation and nationalist identity, which Ole Gilisho allegedly attempted to forge. Entorror was both a place and a defining moment, which many Maasai set against the disharmony and disunity of the present time. The current political struggles over land, resources and power can only be understood in this context.

Appendix I

List of interviewees

Interviews with Maasai elders

Interviews were conducted in Maa, from a written questionnaire, and tape-recorded. They were later translated (by a different assistant, which provided a cross-check) and transcribed into English. Different questionnaires were used for women, prophets, members of the Ole Gilisho family, and Maasai employed by the Delameres at Soysambu. People who described themselves as originally Ol-Aikipiani (pl. Il-Laikipiak, some say Il-Laikipia) absorbed into the Purko section are marked with an asterisk; others insisted they were still Ol-Aikipiani. People's self-description is given in all cases. Interviewee No. 38 (Musungui) claimed to be Ol-Damati (a member of the Il-Damat section) but was believed to be Ol-Toroboni (pl. Il-Torobo). The Il prefix for sections is not given, to save space. For an explanation of age-sets and circumcision groups (RH and LH), see Table 1 (below) and Glossary. Circumcision groups are only given where known.

(a) NAROK. Pilot research in western Narok District, Kenya, summer 1997. By Lotte Hughes with the assistance of David Ole Kenana, Joseph Ole Karia, John Sayiaton and Vincent Ole Ntekerei.

F = Female   L = Laibon/prophet

No.	Name	Section	Age-Set	Place
1	Daudi Ole Teka	Keekonyokie	?RH/Il-Terito	Narok
2	Paul Ole Magiroi	Damat	Il-Terito	Lepisioni
3	Peter Kuyoni Ole Kasura	Purko	RH/Il-Terito	Ilmashariani
4	Karancha Ole Koisikir	Purko	LH/Il-Tareto	nr Ololulunga
5	Sondo Ole Sadera	Purko	Il-Terito	nr Ololulunga
6	Kurao Ole Sadera	Purko	RH/Il-Terito	Ndapupu Olobai
7	Thomas Maitei Ole Motian	Loosekelei	Il-Terito	Olokurto, Mau
8	Ole Ndonyio	Purko	LH/Il-Terito	Naikarra
9	Tarayia Olooloigero	Purko	Il-Nyankusi	Naikarra
10	Kipilosh Ole Kipilosh	Purko	Il-Terito	Naikarra
11	Ole Mantira	Purko	RH/Il-Terito	Entuka nr Uaso Nyiro

(b) LEMEK. Fieldwork October 1999–April 2000 and October 2000, in the Lemek, Aitong and Ololulunga areas of western Narok; Laikipia; Entapipi and Enaiborr-ajijik, Naivasha; and on the Delamere ranch at Soysambu. By Lotte Hughes, with the assistance of Charles Ole Nchoe, Dan Ole Njapit, Martin Ololoigero, Francis Ole Koros, John Ole Karia, Elisabeth Sialala, and briefly Helen Kipetu, John Kimiri and Saiguran Ole Senet. Christopher Chirchir, farm manager at Soysambu, Justice Ole Keiwua and lawyer Keriako Tobiko were interviewed more informally, and are not listed here.

No.	Name	Section	Age-Set	Place
1	Muiya Ole Nchoe	Purko	RH/Il-Terito	Lemek
2	Leperes Ole Gilisho	Purko	RH/Nyankusi	Lemek
3	Kirapusho Ene Gilisho [F]	Purko*	n/a	Lemek
4	Moirori Ole Pusikishu	Laikipiak	RH/Nyankusi	Lemek
5	Toiran Ole Morompi [L]	Purko	LH/Nyankusi	Lemek
6	Iloju Ole Kariankei	Purko*	LH/Nyankusi	Lemek
7	Oloruma (Olkisonkoi Mako) [L]	Purko	?RH/Il-Kitoip	Ol Kimitare
8	Lemasho Ole Morompi [L]	Purko	Il-Kishili	Ndoinyio, Lemek
9	Likwaraa Ole Paswa	Uas Nkishu	LH/Il-Terito	Lemek
10	Samau Ole Kipetu [L]	?Purko	LH/Il-Kitoip	Pardamat
11	Nashilo Ene Liaram [F]	?Purko	n/a	Orkinyei
12	Oloyoogo Olokimolol	Purko	RH/Il-Terito	Orkinyei
13	Olkitojo Ole Sananka	Purko	RH/Nyankusi	Lemek
14	Mapelu Ole Gilisho	Purko	RH/I-Seuri	Ngoswani
15	Nteyo Ole Yiaile	Purko	?RH/Nyankusi	Ngoswani
16	Soitabu Ole Gilisho	Purko	?LH/Nyankusi	Ngoswani
17	Muncheri Ole Nchoe	Purko	?RH/Nyankusi	Ngoswani
18	Nkaburra Ole Njapit	Purko	RH/I-Seuri	Lemek
19	Melempuki Ole Pion	Purko	Il-Nyankusi	Olmusereji
20	Lerionka Ole Ntutu	Purko	RH/Nyankusi	Olchororua
21	Shoriba Ole Gilisho	Purko	?RH/Nyankusi	Nkoilale
22	Parmale Ole Njapit	Purko	?RH/Il-Terito	Nkoilale
23	Salonik Ole Pere	Purko*	RH/Nyankusi	Rumuruti
24	Kireko Ole Kimiri (aka Lenjir)	Purko*	RH/Nyankusi	Rumuruti
25	Kitinti Ole Sadera	Purko	RH/Il-Terito	Engarenarok
26	Nkotumi Ole Kino	Purko	RH/Il-Terito	nr Rumuruti
27	Santayia Ole Ntutu	Purko	LH/Il-Terito	nr Ololulunga
28	Joseph Ole Karia	Purko	LH/I-Seuri	nr Ololulunga
29	Salaton Ole Nchoko	Purko	LH/I-Seuri	Ololulunga
30	Olgiro-Ongu Ole Naurori	Purko	RH/Nyankusi	Oloturoto
31	Pantiya Ene Njapit [F]	Purko	n/a	Ololomeei
32	Lekishon Ole Ketere	Purko	RH/Nyankusi	Ololomeei
33	Kakawua Ole Sengeny	Purko	RH/Nyankusi	Ololomeei
34	Leboo Ole Leintoi	Purko	RH/Il-Terito	Ololomeei
35	Saruni Ole Kipetu	Purko	Il-Kishili	Mararianta
36	Kurito Ene Sengeny [F]	Purko	n/a	Aitong
37	Kapori Ole Kisemei	Purko	RH/Nyankusi	Oloturoto
38	Swahili Musungui	Damat	RH/Nyankusi	Entapipi
39	Nentiyen Ole Kamaamia	Purko	?LH/Nyankusi	Entapipi
40	Kumari Ole Karbolo	Purko	RH/Terito	Entapipi
41	Saiponyari Ole Nchoko	Purko	Il-Nyankusi	Masaantare
42	Olmengo (Ormenko) Ole Nchoko	Purko	Il-Nyankusi	Masaantare
43	Kiter Ene Lemein [F]	Purko	n/a	Oloshapani
44	Ntomoilel Ole Meitaya	Purko	LH/I-Seuri	Ololulunga
45	Salau Ole Gilisho	Purko	RH/Il-Kitoip	Enkerende
46	Ntooto Ene Siololo [F]	Purko	n/a	Enkerende
47	Olochani Ole Karbolo	?Purko	RH/Il-Terito	Enaibor-ajijik
48	Mure Ole Kamaamia	Purko	Il-Nyankusi	Enaibor-ajijik
49	John Ole Karia	Purko*	LH/I-Seuri	Enaibor-ajijik
50	Keloi Ole Gilisho	Purko	LH/I-Seuri	Ngoswani
51	Lendani Ole Sialala	Purko	LH/I-Seuri	Olokirikirai, Mau
52	Reure Ole Sadera	Purko	RH/Il-Kitoip	Soysambu
53	Tuarari Ole Sialala	Purko*	LH/Terito	Soysambu

Interviews with European informants

In Kenya: Desmond Bristow; J.A. 'Jock' Dawson; Lord Delamere

In Britain: Dr W. Plowright, retired veterinary officer, Kenya Colony; Dr Glyn Davies, retired veterinary officer and researcher, was not formally interviewed but supplied much valuable information; Veronica Bellers, great-niece of Arthur Collyer.

Table 1 Maasai age-set chronology from the 1890s

See Glossary for a brief explanation of right-hand and left-hand circumcision groups, and Frans Mol, *Maasai Language, and Culture Dictionary* (Limuru, Kenya: Kolbe Press, 1996), pp. 12–15, for a fuller list from 1755. The following spellings are from Mol; alternatives are given in brackets. Nicknames (not given) are also used. Different sections have different names for some groups and sets; the Purko use the following.

RH Circumcision	LH Circumcision	Age-Set	Dates as Warriors
Il-Ngarbut	Il-Kiponi	Il-Talala	1881–1905
Il-Mirisho	I-Lemek	Il-Tuati (Dwati) II	1896–1917
Il-Meiruturut	Il-Kitoip	Il-Tareto	1911–1929
Il-Tiyieki (Dieki)	Il-Gecherei (Ganchere)	Il-Terito	1926–1948
Il-Kalikal (Kankali)	Il-Kamaniki	Il-Nyankusi II	1942–1959
Il-Terekeyiani	Il-Tiyiogo (Tiyogoni)	I-Seuri	1957–1975
I-Rambauni	Il-Kiropi	Il-Kitoip	1973–1985
Il-Kishili (Kipati, Kisaruni)	Il-Mejooli	Il-Kishiru	1983–1996
			Warriors banned

Appendix II

Chronology of events 1895–1918

1895	British East Africa (East Africa Protectorate) established
November	Kedong massacre, Kedong Valley
1901	Uganda Railway completed
1903	First major land grant to Lord Delamere in the Rift Valley
	CO approves 500-square-mile grant to East Africa Syndicate
1904	
March	Commissioner Sir Charles Eliot offers to resign over land grants
June	Eliot leaves BEA
August	First Maasai Agreement signed with Commissioner Sir Donald Stewart; Maasai agree to leave Rift Valley for two reserves
1905	
spring?	Maasai begin to move to Laikipia and the Southern Reserve
1 April	CO takes over responsibility for BEA from FO
1910	
February	Olonana allegedly asks Gov. Girouard to unite Maasai in one reserve
	1000 Maasai, with 10 000 cattle, move to Southern Reserve for *eunoto* ceremony – effectively the start of the second Maasai move
	Maasai leaders on Laikipia agree to move everyone south
March	Gilbert Murray sends Leys's protest letter about Maasai move to CO
April	CO telegrams Girouard, ordering him to stop the move
30 May	Another Agreement signed by Maasai but not Girouard; never implemented
1911	
7 March	Olonana dies, after allegedly saying Maasai must move south. Succeeded by son Seggi, aged 13
April	Second Maasai Agreement signed on various dates
29 May	1911 Maasai Agreement approved by CO in telegram
May	Settler Galbraith Cole acquitted of murdering alleged sheep thief
June	Move south starts again
August	Move halted on the Mau; deaths reported
October	A third of the 'northern' Maasai back on Laikipia
9 October	Cole deported to Britain
	Girouard lies to CO about promises to settlers of land on Laikipia
1912	
February	Ole Gilisho says reserve is unsuitable and refuses to move; Masikonde agrees, other leaders do not

17 March	Leys claims Ole Gilisho has been threatened with deportation
8 May	CO says the move can go ahead
	Same month, Ole Gilisho starts moving south; consults a lawyer
28 June	Lawyers for the Maasai tell the CO they are taking legal action
July	Girouard leaves BEA after resigning
December?	Leys leaves BEA for Britain in disgrace

1913

27 March	Second Maasai move completed; 10,064 people had moved south since 10 June 1912
May	Leys sails for Nyasaland after home leave
26 May	Maasai Case dismissed by High Court of BEA
December	Appeal dismissed by Court of Appeal for Eastern Africa

1915

| June | Purko warriors attempt to return to Laikipia |

1916   Western area of Maasai Reserve placed in quarantine, after severe outbreaks of BPP and other stock diseases
Attempt to conscript warriors for war effort fails

1917   Unrest reported among Purko and Loitai warriors; ringleaders arrested

1918	The British, tired of Seggi, abolish the post of Paramount Chief
from April	Renewed attempt to conscript warriors for the war effort
11 Sept	Ololulunga 'massacre'
	Lord Delamere protests; offers to mediate between warriors and government.

Notes

Glossary of Maasai Words

1 Frans Mol, *Maasai Language and Culture Dictionary* (Limuru, Kenya: Kolbe Press, 1996).

Preface

1 From an unpublished 'Concept Paper for the Facilitation of Activities towards Institution of the Maasai Case', written by members of the community group SIMOO (Simba Maasai Outreach Organisation), Kenya, 2002. Copy supplied to author.
2 Memorandum by Jackson, 15 August 1903, published in 'Correspondence Relating to the Resignation of Sir Charles Eliot, and to the Concession to the East African Syndicate', Cd. 2099, Africa No. 8 (London: HMSO, 1904), p. 11. Jackson to Lansdowne, 22 Feb. 1904, Desp. 137, 8 Mar. 1904, FO 2/836; also in Cd. 2099, *idem*, p. 6.
3 Narok District A/R 1928, p. 5.

Part I The moves and what led up to them

Chapter 1 Introduction

1 Eliot to Lansdowne, FO 2/835, PRO.
2 Leys to Harvey, No. 3, 20 May 1911, Harvey Letters, held by the author. These are to be deposited in a public archive.
3 This is not the term people use to describe themselves. For a recent discussion of 'Il-Torobo', see Lee Cronk, *From Mukogodo to Maasai: Ethnicity and Cultural Change in Kenya* (Boulder, CO: Westview Press, 2004).
4 Some examples of fashion coverage featuring Maasai include the photoshoot by David Bailey in *Vogue*, British edition, January 2004; *Harpers & Queen*, March 2002; 'Bailey's Africa' and 'Safaris for all Seasons', *Daily Telegraph*, 22 March 2003.
5 Isaac Sindiga, 'Land and population problems in Kajiado and Narok, Kenya', *African Studies Review*, 27, No. 1 (March 1984), 26; Thomas Spear, Introduction, T. Spear and R. Waller (eds), *Being Maasai: Ethnicity and Identity in East Africa* (Oxford: John Currey, 1993).
6 Marcel Rutten, *Selling Wealth to Buy Poverty: The Process of the Individualization of Landownership Among the Maasai Pastoralists of Kajiado District, Kenya, 1890–1990* (Saarbrücken and Fort Lauderdale: Verlag Breitenbach, 1992), p. 6.
7 For example, John L. Berntsen, 'Pastoralism, Raiding and Prophets: Maasailand in the Nineteenth Century', Ph.D. thesis, Wisconsin (1979); Richard D. Waller, 'The Lords of East Africa: The Maasai in the mid-Nineteenth

188

Century (*c.* 1840–1885)', Ph.D. thesis, Cambridge (1979); Alan Jacobs, 'The Traditional Political Organisation of the Pastoral Masai', Ph.D. thesis, Oxford (1965); published works by all these authors; John Galaty's chapters in Spear and Waller, *Being Maasai,* and Galaty and P. Bonte (eds), *Herders, Warriors and Traders: Pastoralism in Africa* (Boulder, CO: Westview, 1991); and others.

 8 See my D. Phil. thesis, 'Moving the Maasai: A Colonial Misadventure' (Oxford, 2002). For length reasons I have cut the section on comparative resistance from this book.

 9 M. P. K. Sorrenson, *Origins of European Settlement in Kenya* (Oxford: OUP, 1968), p. 276.

10 *KLC Evidence and Memoranda,* Vol. 2 (London: HMSO, 1934), p. 1223. It listed placenames derived from Maa, described by Maasai witnesses as 'conclusive proof of the fact that all these regions were ours and beneficially occupied and populated by us'.

11 Joseph Thomson, *Through Masai Land* (London: Sampson Low, Marston, Searle and Rivington, 1885). There were several subsequent editions and translations; all quotes will be from the first unless otherwise stated. Thomson's influence is further discussed in my chapter in G. de Vos, L. Romanucci-Ross and Takeyuki Tsuda (eds), *Ethnic Identity: Problems and Prospects for the Twenty-First Century,* 4th edn (Walnut Creek, CA: AltaMira Press, 2006).

12 There were precedents for East Africans employing European lawyers to pursue high court actions, but these were often defensive. See Justin Willis, 'Killing Bwana: Peasant revenge and political panic in early colonial Ankole', *JAH,* 35 (1994).

13 Norman Leys, *Kenya* (London: The Hogarth Press, 1924), p. 63 of 2nd edn (1925). All quotes will be from the latter unless otherwise stated.

14 For space reasons, many footnotes have been cut. See my dissertation for more information.

15 Dan Brockington, *Fortress Conservation: The Preservation of the Mkomazi Game Reserve, Tanzania* (Oxford: James Currey, 2002).

16 John L. Berntsen, 'Maasai age-sets and prophetic leadership, 1850–1910', *Africa,* 49, No. 2 (1979), 145.

17 R. L. Tignor, *The Colonial Transformation of Kenya: The Kamba, Kikuyu and Maasai from 1900 to 1939* (Princeton: Princeton University Press, 1976); 'The Maasai warriors: Pattern maintenance and violence in colonial Kenya', *JAH,* 13 (1972).

18 Jan Vansina, *Oral Tradition as History* (Oxford: James Currey, 1997), p. 68.

19 Jacobs does not define Maasai traditional history, but says it probably began about 1778 with the birth of the prophet Subet (Supeet), 'Pastoral Masai', p. 54. Waller defines traditions as 'an abridgement of the past ... structured in such a way as to express enduring social ideas in a historical framework', 'Lords of East Africa', p. 402.

20 Greg Dening, *Performances* (Chicago: University of Chicago Press, 1996), p. 50.

21 For example, the current Lord Delamere behaves as if he knows very well that he is a social anachronism in the twenty-first century, and plays up to that in interview. As the descendant of settlers who despised central government and often defied it, he offers a view that is no less oppositional and subversive than that of the Maasai, in its own way.

22 E. Tonkin, *Narrating our Pasts: The Social Construction of Oral History* (Cambridge: Cambridge University Press, 1992), p. 8.

23 James Scott, *Domination and the Arts of Resistance: Hidden Transcripts* (New Haven, CT: Yale University Press, 1990). He defines public transcripts as 'a shorthand way of describing the open interaction between subordinates and those who dominate', p. 2.

24 Diana S. Wylie, 'Critics of colonial policy in Kenya, with special reference to Norman Leys and W. McGregor Ross', M.Litt. thesis, Edinburgh (1974); Wylie, 'Norman Leys and McGregor Ross: A case study in the conscience of African empire, 1900–39', *Journal of Imperial and Commonwealth History*, 5, No. 3 (May 1997); John W. Cell (ed.), *By Kenya Possessed: The Correspondence of Norman Leys and J. H. Oldham, 1918–1926* (Chicago: University of Chicago Press, 1976).

25 In 1930 Leys applied for reinstatement to the colonial service, but was rebuffed. Leys to Secretary of State, 25 July 1930, Murray Papers, Box 32, Bodleian Library, Oxford.

26 Cell, *Possessed*, note p. 322.

27 T. H. R. Cashmore, 'Studies in District Administration in the East African Protectorate, 1895–1918', Ph.D. thesis, Cambridge (1965); I consulted an earlier draft, see Bibliography. G. H. Mungeam, *British Rule in Kenya, 1885–1912* (Oxford: Clarendon Press, 1966). G. R. Sandford, *An Administrative and Political History of the Masai Reserve* (London: Waterlow & Sons, 1919). Sorrenson, *Origins*; Tignor, *Colonial Transformation*.

28 Leys, *Kenya*; W. McGregor Ross, *Kenya from Within: A Short Political History* (London: George Allen & Unwin, 1927).

29 Sandford, *Administrative History*, pp. 2, 3, 28, 58. The Eliot quote is from the Introduction, dated 14 Nov. 1904, to A. C. Hollis, *The Masai: Their Language and Folklore* (Oxford: Clarendon Press, 1905), p. xvii.

30 For example, Gabriele Somer and Rainer Vossen, 'Dialects, sectiolects, or simply lects? The Maa language in time perspective' in Spear and Waller, *Being Maasai*, lists 22, p. 30; Rutten, *Selling Wealth*, lists 14, p. 133.

31 See chronologies in Mol, *Maasai Language*, pp. 12–15; Jacobs, 'Traditional Political Organisation', p. 49; H. Fosbrooke, 'The Masai age-group system as a guide to tribal chronology', *African Studies*, 15 (1956); and my Appendix 1.

32 Dorothy L. Hodgson 'Pastoralism, patriarchy and history: Changing gender relations among Maasai in Tanganyika, 1890–1940', *JAH*, 40 No. 1 (1999); Hodgson (ed.), *Rethinking Pastoralism in Africa: Gender, Culture and the Myth of the Patriarchal Pastoralist* (Oxford: James Currey, 2000).

33 For example, see Spear and Waller, *Being Maasai*; John L. Berntsen, 'The enemy is us: Eponymy in the historiography of the Maasai', *History in Africa*, 7 (1980).

34 See Sorrenson, *Origins*, Part 1, pp. 9–58.

35 B. Berman and J. Lonsdale, *Unhappy Valley: Conflict in Kenya & Africa, Book 1: State and Class* (Oxford: James Currey, 1992), p. 1.

36 Sorrenson, *Origins*, p. 27. There was some early white settlement – around 140 planters by 1914 and 220 large European estates by 1920, R. M. A. Van Zwanenberg with Anne King, *An Economic History of Kenya and Uganda, 1800–1970* (London and Basingstoke: Macmillan, 1975), pp. 61–3 and seq. This economy collapsed when cotton prices crashed.

37 Letter dated 1 April 1930, in Delamere, 'Letters from Kenya', Mss. Afr. s. 1424 (1), RHO, f102.
38 'Report by HM Commissioner on the East Africa Protectorate, April 1903', Cd. 1626, reproduced in G. H. Mungeam (ed.), *Select Historical Documents, 1884–1923* (Nairobi: East African Publishing House, 1978), p. 94.
39 Berman and Lonsdale, *Unhappy Valley*, 1, pp. 2, 3, 6, 34–6.
40 Wylie, 'Critics', p. 38. But she does not say whether they were already acquainted.
41 For example, Leys to Mrs Ross, 19 August 1923, 7 Sept. 1923, Box 15, McGregor Ross Papers, RHO. Also Leys to Harris, 27 Nov. 1926, ASAPS Papers, RHO.
42 Wylie, 'Critics', Abstract.
43 Ibid., pp. 42, 45. When his paternal grandfather refused to allow his grandsons to return to their father John, the latter launched legal action, 'famous at the time', which led to his imprisonment for contempt of court. Wylie suggests that both Leys's puritanical roots and US sojourn, at a New England school founded by an evangelist, fundamentally influenced his later political activities.
44 This information is from Wylie. But the Protectorate Blue Book for 1912 states that Leys was appointed on 30 September 1904.
45 For example, in an appendix to his 1909 health report on Nakuru, Leys alerted the authorities to a 'pandemic' of VD among Africans (linked to prostitution and labour patterns along the railway) and said this could only be checked by a radical change in living conditions. He was ahead of his time in moving beyond a purely medical model of health. See Tignor, *Transformation*, p. 180; the original report is in PC/NZA 2/3, KNA.
46 Leys, *Kenya* (1925 edn), p. 326.
47 Ibid., p. 86.
48 Sandford, *Administrative History*, p. 15. The number of fatalities is from Frederick Jackson, *Early Days in East Africa* (London: Edward Arnold, 1930). Other accounts differ. See Ainsworth to Hardinge, 3 Dec. 1895, Enc. in Hardinge to Salisbury, 19 Dec. 1895, FO 107/39 and other despatches in Jan. 1896. R. M. Maxon and D. Javersak describe the contradictions in accounts of this episode in 'The Kedong Massacre and the Dick Affair: A problem in the early colonial historiography of East Africa' *History in Africa*, 8 (1981).
49 Leys, *Kenya*, pp. 76, 94.
50 Ibid., p. 104.
51 Personal communication with Veronica Bellers, Collyer's great-niece.
52 Leys, *Kenya*, p. 135.
53 Ibid., pp. 137–8.
54 Quotes are from Wylie, 'Critics', p. 35.
55 Anonymous obituary, *The Friend* (London: Religious Society of Friends, 9 Feb. 1940), p. 80.
56 Wylie, 'Critics', p. 168.
57 See Ch. 7 of my thesis for a fuller description of his life; cut for length reasons.
58 References in Jackson, *Early Days*, include pp. 293–4, 329–30.
59 I am indebted to Frans Mol for sharing with me his unpublished notes on the life of Ole Gilisho. The passages quoted here are taken from these.

Chapter 2   The moves

1   Peter Rigby, *Persistent Pastoralists: Nomadic Societies in Transition* (London: Zed Books, 1985), p. 67. His translation differs slightly.

2   Joseph Thomson, *Through Masai Land* (London: Sampson Low, Marston, Searle & Rivington, 1885), pp. 407, 408.

3   F. D. Lugard, *The Rise of our East African Empire* (Edinburgh: William Blackwood & Sons, 1893), p. 419.

4   C. Eliot, *The East Africa Protectorate* (London: Edward Arnold, 1905), p. 3.

5   G. R. Sandford, *An Administrative and Political History of the Masai Reserve* (London: Waterlow & Sons, 1919), p. 20. This is based on A. C. Hollis, 'Memorandum on the Masai', 5 July 1910, published in 'Correspondence Relating to the Masai' (London: HMSO, Cd. 5584, June 1911), p. 15.

6   H. H. Johnston, *The Kilima-Njaro Expedition: A Record of Scientific Exploration in Eastern Equatorial Africa* (London: Kegan Paul & Trench, 1886), pp. 537, 552. See also Ch. 1 of Johnston, *The Uganda Protectorate* (London: Hutchinson, 1902) for glowing descriptions of 'empty' country in its eastern province (soon to be absorbed into BEA) that reminded him of Scotland, Wales, Surrey and Sussex. In *The Nineteenth Century and After* magazine, in October 1908, he advocated moving the Maasai to make way for white settlement on Laikipia.

7   Ainsworth to Craufurd, 13 June 1899, f212 of No. 68 in Machakos A/Rs, DC/MKS (cover had fallen off, hence this file was not clearly labelled), KNA.

8   Eliot, 'Report on the East Africa Protectorate', Cd. 769 (1901), p. 8. The reference to Indian settlement is in Eliot to Lansdowne, 5 Jan. 1902, FO 2/569.

9   M. P. K. Sorrenson, *Origins of European Settlement in Kenya* (Nairobi: OUP, 1968), p. 47.

10   Norman Leys, *Kenya* (London: The Hogarth Press, 1931), pp. 79–80.

11   B. Berman, *Control & Crisis in Colonial Kenya* (Oxford: James Currey, 1990), pp. 150–1.

12   C. C. Wrigley, Ch. V, in V. Harlow and E. M. Chilver (eds), *History of East Africa*, Vol. 2 (Oxford: OUP, 1965), p. 217. For Zionist settlement plans, see Sorrenson, *Origins*, Ch. 2.

13   B. Berman and J. Lonsdale, *Unhappy Valley: Conflict in Kenya & Africa, Book 1* (Oxford: James Currey, 1992), p. 89; Sorrenson, *Origins*, p. 146.

14   Some prophecies are described in C. H. Stigand, *The Land of Zinj* (London: Constable, 1913).

15   E. Huxley, *White Man's Country: Lord Delamere and the Making of Kenya*, Vol. 1 (London: Macmillan, 1935), p. 104. She claims this was the only grant he received from government; the rest he bought from other settlers.

16   'Lord Delamere's Emigration Scheme', *The Times* (London: 26 March 1904).

17   Eliot to Lansdowne, 10 Oct. 1903, Enc. D, No. 467, FO 2/846, printed as Cd. 2099, Africa No. 8 (1904), 'Correspondence Relating to the Resignation of Sir Charles Eliot, and to the Concession to the East Africa Syndicate' (London: HMSO, July 1904). See Eliot, *The East Africa Protectorate* (London: Edward Arnold, 1905), pp. 105–6, 170, 310, for his views on reserves.

18   Eliot, 'Memorandum on Native Rights in the Naivasha Province', 7 Sept. 1903, published in Cd. 2099, *idem*.

19   These sentiments, and warnings of 'race trouble' if Maasai were allowed to stay in the Rift, were reiterated in Ainsworth's 1904 'Memorandum for the Land Committee', DC/MKS/26/3/1, KNA.

20 The map, dated February 1904, is in Enc.1, Desp. 495, 22 July 1904, FO 2/838.

21 Jackson to Lansdowne, 22 Feb. 1904, quoted in Desp. 137, Lansdowne to Eliot, 8 Mar. 1904, FO 2/836; also Enc. G in Conf. dated 16 Mar. 1904, FO 2/846; published in Cd. 2099, *idem.*

22 Lansdowne to Eliot, Enc. E, No. 578, 27 Nov. 1903, Appendix B, FO 2/846.

23 Jackson to Lansdowne, 22 Feb. 1904, No. 10 in Cd. 2099, *idem.*

24 Hobley said of the map that 'the red areas are the <u>permanent reserves</u> proposed by Mr Ainsworth and myself' [his underlining], Enc. 3, Desp. 495, Hobley to Lansdowne, 22 July 1904, FO 2/838.

25 Enc. 1 in Desp. 495, 22 July 1904, FO 2/838. Also see C. W. Hobley, *Kenya: From Chartered Company to Crown Colony* (London: H. F. & G. Witherby, 1929). There is a useful account of how Hobley's views on the Maasai differed from Eliot's in Mungeam's introduction to the 2nd edn (London: Frank Cass, 1970), p. xii.

26 See memo to Eliot by Major Harrison of the 3rd KAR and Ainsworth, Enc. 4 in Eliot to Lansdowne, Desp. 351, 20 May 2004, FO 2/836.

27 Introduction to Hobley, *Kenya* (1970 edn), p. xi.

28 'Journey from Naivasha to Baringo and the Laikipia Highlands', 24 June 1904, Enc. 1, Desp. 493, 22 July 1904, FO 2/838; Tel. 124, Hobley to Lansdowne, 1 July 1904, FO 2/842.

29 Chamberlain, a former journalist in Britain and South Africa, was definitely British. He implied that Flemmer was, too, in an undated letter to the press in support of Eliot, writing 'as over-sea Britishers'. Letter headed 'The Colonisation of Africa: Sir Charles Eliot's Case', no addressee, Robert Chamberlain Papers, RHO, f118.

30 For Eliot's defence of his actions, see Cd. 2099 and the preface to Eliot, *Protectorate*. Hollis defended him in his unpublished *Autobiography of Alfred Claud Hollis*, RHO. Sorrenson, *Origins*, Ch. 4, covers this episode in detail.

31 Desp. No. 217, Eliot to Lansdowne, 5 Apr. 1904, in FO 2/835. The worst of his attack on Jackson was cut from this despatch for publication as No. 23 in Cd. 2099.

32 Desp. No. 234, Eliot to Lansdowne, 9 Apr. 1904, FO 2/835; Sorrenson, *Origins*, p. 76.

33 Tel. 33, Eliot to Lansdowne, 3 Feb. 1904, FO 2/842.

34 Chamberlain to Eliot, 12 Mar. 1904, f79, and 21 June 1904, f110; Chamberlain to Milner, 29 July 1904, ff149–57; undated telegram in June 1904, Chamberlain to Eliot, ff107–8, in response to Eliot's telegram telling him of his resignation, Chamberlain Papers, RHO.

35 Sandford, *Administrative History*, p. 24; Hollis, memo, 'Correspondence Relating to the Masai', p. 16.

36 Under Secretary of State to Chamberlain, 6 Feb. 1906, f322, Chamberlain Papers, RHO.

37 Stewart to Lansdowne, 4 Oct. 1904, Desp. 619, FO 2/838.

38 Chamberlain to Elgin, 11 Mar. 1906, f345, Chamberlain Papers, RHO.

39 '... it is rather unlucky that he should have hurried matters so much,' minute on Tel. 146A, Stewart to Lansdowne, 16 Aug. 1904, FO 2/842.

40 Hollis, *Autobiography*, Vol. 3, p. 45, RHO. Hollis was an Assistant Collector in BEA from 1897, then Secretary to the Administration and Private Secretary to Eliot from 1901, before becoming Secretary for Native Affairs from 1907.

41 The meeting is described in Hill's memo on the 'Masai Question', 19 Aug. 1904, FO 2/842.

42 Stewart to CO, 'Memorandum on the Removal of the Masai from the Rift Valley and the Vicinity of Nairobi', Enc. 1 in No. 119, 3 Feb. 1905, 'Further Correspondence Respecting East Africa January to March 1905', p. 193. Also a report of an expedition against the Sotik mentions that warriors were away moving stock to Laikipia, Desp. 327, Stewart to Lansdowne, 8 June 1905, CO 533/2.

43 Sandford, *Administrative History*, p. 25. Taken from Hollis, 'Memorandum on the Masai', 5 July 1910, in 'Correspondence Relating to the Masai', p. 14.

44 Leys, *Kenya*, p. 102.

45 Undated anonymous letter, no source, signed 'A well wisher and admirer', ASAPS Papers, Mss. Brit. Emp. s. 22, G131, RHO.

46 Gerald Hanley, *Warriors and Strangers* (London: Hamish Hamilton, 1971), pp. 293–5.

47 J. Ford, *The Role of the Trypanosomiases in African Ecology: A Study of the Tsetse Fly Problem* (Oxford: Clarendon Press, 1971), p. 139. Also see R. D. Waller, '*Emutai*: Crisis and response in Maasailand, 1883–1902', in D. Johnson and D. M. Anderson (eds), *The Ecology of Survival* (London: Lester Crook, 1988), p. 101; Waller, 'The Maasai and the British 1895–1905: The origins of an alliance', *JAH*, 17, No. 4 (1976), 530–3.

48 Bagge ordered them to go. No human figures were given, but they took with them 10,000 cattle and 30,000 small stock, Sandford, *Administrative History*, p. 27.

49 R. D. Waller, 'The Lords of East Africa: The Maasai in the Mid-Nineteenth Century (*c.* 1840–1855)', Ph.D. thesis (Cambridge: 1978), p. 56.

50 Waller, 'Origins', p. 540; '*Emutai*', pp. 80, 92–3, 107.

51 Leys, *Kenya*, p. 102. The 1906 figure is in the Laikipia District A/R, the 1911 figure in Sandford, *Administrative History*, p. 3. These were probably gross underestimates.

52 Waller, 'Lords of East Africa', p. 45.

53 Laikipia District Survey 1906–11, p. 4. In the A/R for the year ending March 1913 the 1906 figure was raised to 1.77m, and current numbers of sheep put at ?844,000 (microfiche illegible), a deficit of nearly one million. DC McClure wrote that sheep numbers had probably been overestimated in 1906 and underestimated in 1912–13 by between 300,000 and 600,000. The ratio of cattle to sheep was one to four.

54 T. H. R. Cashmore, 'Your Obedient Servants', p. 340 (see Bibliography); Laikipia District Survey, pp. 7–8, 10.

55 Waller, 'Origins', p. 550; '*Emutai*', pp. 109–10.

56 Laikipia District Survey, pp. 9–10. Repeated in slightly different wording in Enc. 1, Desp. 14, Belfield to Harcourt, 6 Feb. 1913, CO 533/116.

57 Quarterly Report for Laikipia ending March 1910.

58 Olonana reportedly told government that he favoured reunification in one reserve in December 1908, Laikipia District Survey, p. 6. He repeated the request in January 1909 when Hayes Sadler met him at Ngong, and later to Girouard in February 1910.

59 R. A. I. Norval, B. D. Perry and A. S. Young, *The Epidemiology of Theileriosis in Africa* (London: Academic Press, 1992), p. 48.

60 See P. F. Cranefield, *Science and Empire: ECF in Rhodesia and the Transvaal* (Cambridge: Cambridge University Press, 1991).

61 Winston Churchill, *My African Journey* (London: Hodder & Stoughton, 1908), p. 40.

62 Girouard to CO, 1 and 4 Apr. 1910, Desp. 175, CO 533/71.

63 Laikipia District A/R, year ended 31 March 1911.

64 Sorrenson, *Origins*, p. 126. Delamere was not an original applicant, but had bought the Thorne brothers' claim on Laikipia by November 1905, fn 33, p. 106.

65 All biographical information about Girouard is taken from an unpublished biography, *The Lily and the Rose*, by Michael L. Smith which the author has given his permission to quote. My thanks to Tony Kirk-Greene for showing me this.

66 Ibid., p. 294, quoting Girouard's letter to his father Désiré, 22 July 1907.

67 Ibid., p. 322, quoting from Girouard's unpublished work 'The Imperial Ideal'; also pp. 305, 312.

68 Sorrenson, *Origins*, p. 126.

69 Girouard to Read, 3 Mar. 1910, CO 533/72; report by Hollis and Collyer of Girouard's meeting with Olonana on 2 February, Enc. 2, 4 Feb. 1910, in Desp. 14, CO 533/116.

70 Leys to Murray, 3 Feb. 1910, Mss. 148, Murray Papers, Bodleian Library, Oxford. See also Jackson to Crewe, 7 Mar. 1910, CO 533/72. Civil servant George Fiddes noted that the subject could 'easily give rise to a tornado in the House of Commons' if word got out.

71 Crewe to Girouard, Tel. of 19 Apr. 1910 followed by full despatch 22 Apr., CO 533/72; Sorrenson, *Origins*, pp. 128, 200–1.

72 Enc. 3 in Conf. Desp. 14, Belfield to Harcourt, 6 Feb. 1913, CO 533/116.

73 Laikipia District Survey 1906–11, p. 13; Sandford, *Administrative History*, p. 30.

74 Laikipia District Survey 1906–11, p. 14.

75 Summary in ibid., p. 14. The full report and associated letters are in Enc. 5, Desp. 14, Belfield to Harcourt, 6 Feb. 1913, CO 533/116.

76 Ibid., Enc. 1 in Desp. 14. The copy on file is undated.

77 These 1910 reports by Collyer were sent to the CO in February 1913, only after Harcourt had wired BEA demanding information he suspected was being withheld; ibid., all enclosures in Desp. 14. An angry note in red ink by Harcourt, dated 11 March 1913, said: 'I cannot understand Mr Collyer's report of 29/8/10 never having been sent home to us when we were enquiring as to the attitude of the Masai to the move. I now think that the questions and debate in the H of C in 1911 must have been inspired by a knowledge of this report.'

78 For more information on Harvey, see his obituary in *The Friend* (13 May 1955), pp. 491–3.

79 H. W. Nevinson, *A Modern Slavery* (London and New York: Harper, 1906).

80 Initially called *The Anti-Slavery Reporter and Aborigines' Friend*, Vols 1–2 (1909–13), RHO.

81 John W. Cell, Introduction, in Cell (ed.), *By Kenya Possessed: The Correspondence of Norman Leys and J. H. Oldham, 1918–1926* (Chicago: University of Chicago Press, 1976), p. 17.

82 Leys to Harvey, No. 18, 22 Mar. 1912, Harvey Letters; Leys to Travers Buxton, 26 Apr. 1913, ASAPS Papers, G131, RHO. See Diana S. Wylie, 'Critics of

Colonial Policy in Kenya, with Special Reference to Norman Leys and W. McGregor Ross', M. Litt. thesis (Edinburgh: 1974), for a full description of these interlinking networks.

83 Leys to MacDonald, copied to Harvey, Murray, No. 26, 17 July 1912, Harvey Letters.

84 Girouard to Morel, 1 Oct. 1911, Morel Papers, LSE. Also in Girouard to Harcourt, 30 Sept. 1911, CO 879/112.

85 Leys to Harvey, No. 1, 17 Oct. 1910, Harvey Letters.

86 Cashmore, 'Obedient', Vol. 2, p. 347. An Oxford graduate, he arrived in BEA in 1902 at the age of 22.

87 Leys to Murray, 30 Apr. 1910, Murray Papers, Bodleian.

88 Ross to mother, 21 Aug. 1908, Ross Papers, RHO.

89 The interview was on 3 June, described the same day in Leys to Murray, 3 June 1910, Murray Papers (148, ff92–6), Bodleian.

90 Ross to mother, No. 439, 24 May 1910, Ross Papers.

91 Ross to mother, No. 490, 30 May 1910, Ross Papers.

92 Leys to Harvey, No. 3, 20 May 1911, Harvey Letters.

93 Enc. 1 in Girouard to Harcourt 18 Apr. 1911, CO 533/116. The original treaty has never been found.

94 Minute by Harcourt on Girouard to Harcourt, 15 Mar. 1911, CO 533/85.

95 'Report by Crewe-Read on the Death of Lenana', 8 Mar. 1911, Southern Masai Reserve District Records 1908–11, DC/KAJ.1/1/1, KNA.

96 MacDonald to Harcourt, 18 May 1911, CO 533/86.

97 *Hansard*, 5th series, Vol. 28, debate of 20 July 1911, cols 1324–1353.

98 Charles Miller, *The Lunatic Express* (New York: Macdonald, 1971), p. 495.

99 Rumuruti (Laikipia) District A/R, year ending 31 Mar. 1912, LKA/1, KNA, p. 1, with cover note by H. B. Popplewell. The total head of stock still on Laikipia at the end of July was about 50,000 cattle and a 'proportionate number of sheep'. By the year's end, cattle numbers had gone back up to 160,647 and sheep to 1,068,100.

100 Ibid., p. 2.

101 Girouard to Harcourt, No. 497, 6 Sept. 1911, CO 533/90.

102 ASAPS Papers, RHO; copies also in the Harvey Letters. The first letter is undated and unsourced. The third, dated 7 January 1912, has a Swiss post-mark; four sepia photographs are enclosed, one clearly showing a woman's skeleton. The report quoted is from the second letter, dated 25 September 1911, London.

103 Wylie, 'Critics', pp. 75–6.

104 Leys to Harvey, MacDonald, Murray, No. 12, 7 Oct. 1911, Harvey Letters.

105 For Boedeker, sometimes spelled Bődeker, see Leys to Harvey, MacDonald, No. 8, 3 Sept. 1911. The other reports are in Enc. 1 and 2, No. 497, 6 Sept. 1911, CO 533/90.

106 Leys to Harvey et al., ibid., No. 12.

107 Leys to Harvey, MacDonald, No. 6, 22 Aug. 1911.

108 Sandford, *Administrative History*, p. 32. He lifts Girouard almost word for word from Girouard to Harcourt, Desp. 548, 30 Sept. 1911, CO 533/90, pp. 8–9.

109 His description of the soldiers' origins as Sudan and Germany may be a reference to Nyamwezi from German East Africa and Nubians from the Sudan. For information on Ole Mootian, see K. King, 'The Kenya Maasai and the

protest phenomenon, 1900–1960', *JAH*, 12 (1971); King and A. Salim (eds), *Kenya Historical Biographies* (Nairobi: East African Publishing House, 1971); Waller, Ch. 11 in Spear and Waller, *Being Maasai*.

110   The Harcourt quote is a minute on Girouard to Harcourt, Tel. 190, 5 Sept. 1911, CO 533/90. Girouard said the move had been made 'quite voluntarily'. Sandford lifted Girouard's wording from this and despatch No. 497 dated one day later.

Chapter 3   In search of the truth

1   Leys to Harvey, No. 3, 20 May 1911, Harvey Letters.

2   DC Eldama Ravine to PC Naivasha, 25 Sept. 1911, PC/RVP/6E/1/1, KNA.

3   Diary of Tour by E. C. Crewe-Read, 19 Sept. 1911, ibid.

4   Some individual affidavits, for example Ole Nchoko's, stated there were human deaths while moving 'owing to the lack of food and water for the cattle and scarcity of water and milk for women and children and food for the men and also to the cold on the Mau Hills'. No numbers given. Copy in G131 of the ASAPS Papers, RHO.

5   Leys to Harvey, No. 12, 7 Oct. 1911, Harvey Letters.

6   *Hansard*, 5th series, Vol. 37, 16 Apr. 1912, cols 180–1. Previous reference is *Hansard*, 5th series, Vol. 28, debate of 20 July 1911, cols 1324–53.

7   Not listed among my interviewees, since I only spoke to her briefly post-fieldwork in January 2003.

8   Leys to Harvey, No. 9, 10 Sept. 1911, Harvey Letters; Leys did not name him, but Ronald Donald is listed as town magistrate in the Blue Book 1910–11, though there were staff changes that year. He also attended a major meeting to discuss the move on 5 Dec. 1912 in Nakuru (CO 533/109).

9   Girouard to Harcourt, Desp. 548, 30 Sept. 1911, CO 533/90, pp. 13, 15.

10   Allgeyer to Girouard, 6 Sept. 1910; Peel to Girouard, 19 Oct. 1910, enc. in Desp. 330, 15 June 1911, CO 533/88.

11   F. H. Goldsmith, *John Ainsworth: Pioneer Kenya Administrator, 1864–1946, being the hitherto unpublished memoirs of Col. John D. Ainsworth* (London: Macmillan, 1955), pp. 82–5. 'Giddy Masai' presumably comes from Kipling: 'some share our tucker with tigers, / and some with the gentle Masai / (Dear boys!), / Take tea with the giddy Masai', *The Lost Legion* (1895).

12   MacDonald's report is an enc. in Desp. 624, Girouard to Harcourt, 3 Nov. 1911, CO 533/92. Ross to mother, No. 516, no date, McGregor Ross Papers, RHO.

13   Minutes on Desp. 624, *idem*, and on Tel. 264, Girouard to Harcourt, 25 Nov. 1911, CO 533/92.

14   Minute on Girouard to Harcourt, Desp. 652, 21 Nov. 1911, CO 533/92.

15   Browne to Pickford, 27 Aug. 1911, Hollis to Ainsworth 11 Sept. 1911, PC/RVP/6E/1/1, KNA; Leys to Harvey, No. 23, 29 May 1912. Enclosed with the Harvey Letters is an undated hand-drawn map on yellow parchment, marking areas where four species of tsetse fly were to be found in the Southern Reserve. The author (the handwriting is not Leys's) wrote on it in pencil: 'Trans Amala country that the Masai are being sent to very waterless and nearest water infected by fly probably palpalis.' Leys sent the map to Harvey in 1910. See Waller, 'Tsetse fly in western Narok, Kenya', *JAH*, 31 (1990) for the history of tsetse in this area.

16 G. R. Sandford, *An Administrative and Political History of the Masai Reserve* (London: Waterlow & Sons, 1919), p. 33.
17 Enc. in Desp. 45, Bowring to Sec. of State, 5 July 1912, CO 879/112.
18 Leys to Harvey, MacDonald, No. 13, 5 Nov. 1911, Harvey Letters; MacDonald to Harcourt, 27 Nov. 1911, CO 533/92.
19 Enc. 6 in Desp. 14, 6 Feb. 1913, CO 533/116.
20 Collyer to McClellan, 12 Apr. 1912, Enc. 9 in Desp. 14, CO 533/116.
21 Leys to Harvey, Nos 17 and 19, 17 and 27 Mar. 1912, Harvey Letters; references to the threat are in Desp. 14, *idem*, CO 533/116; Norman Leys, *The Colour Bar in East Africa* (London: Hogarth Press, 1941), p. 25. He does not refer to himself by name.
22 In October 1911 C. R. W. Lane, PC Naivasha, told Harcourt face to face that he thought promises of land had in fact been made. When challenged, Girouard wired a denial. Harcourt minuted his doubts about Girouard on 11 Oct. 1911, CO 533/91.
23 Morrison was a graduate of Aberdeen University, who was called to the Bar in 1903, moved to East Africa the following year as a magistrate and went into private practice before 1910, Cashmore, 'Obedient', fn pp. 376, 377.
24 Reports of Criminal Case No. 94 of 1911 at Nakuru Magistrates Court (sitting as a High Court it became Session Case No. 25 of 1911) and Cashmore (p. 365), say the victim, Siongo wa Nasuru, was Kikuyu; this was contradicted by Arthur Cole in personal correspondence. For analyses of the case, see Robert M. Maxon, 'Judgement on a colonial governor: Sir Percy Girouard in Kenya', *Transafrican Journal of History*, Vol. 18 (1989); and Michael L. Smith, *The Lily and the Rose* (unpublished biography), Ch. 30.
25 Ross to mother, No. 515, 17 Sept. 1911, McGregor Ross Papers, RHO.
26 The original, and amended, deportation orders are in Girouard to Harcourt, No. 77, 15 Sept. 1911, CO 533/90 and Conf. 93, 5 Oct. 1911, CO 533/91. See the first of these registers for the heated correspondence between the two over the deportation. For his wife's account of their early years, see Eleanor Cole, *Random Recollections of a Pioneer Kenya Settler* (Suffolk: Baron Publishing, 1975). She wrote nothing about this episode, having only met Cole in 1916. For more stories of Galbraith and his brother Berkeley, including a warm endorsement of Galbraith's actions, see Errol Trzebinski's *Silence Will Speak* (London: Heinemann, 1977) and *The Kenya Pioneers* (London: Heinemann, 1985).
27 Tel. of 5 Sept. 1911, Girouard to Harcourt, minuted 'This is unpardonable' at CO, and the reply, CO 533/90.
28 Leys to Harvey, No. 4, 11 Aug. 1911, Harvey Letters.
29 Ross to mother, No. 511, 19 Aug. 1911, McGregor Ross Papers, RHO.
30 *Hansard*, 5th series, Vol. 28, 12 July and 20 July 1911; no exact date given for the *Leader* story in September, T. H. R. Cashmore, 'Your Obedient Servants' (see Bibliography), Vol. 2, p. 368.
31 Leys to Harvey and MacDonald, No. 10, 16 Sept. 1911, Harvey Letters.
32 Quoted in the *Leader*, September. No original date given.
33 Leys to Harvey, No. 10, 16 Sept. 1911, Harvey Letters.
34 Introduction to 4th edn, Leys, *Kenya*, pp. ix, x.
35 Leys to Harvey, No. 11, 19 Sept. 1911, Harvey Letters. For settler opinion, see Lord Hindlip, *British East Africa: Past, Present and Future* (London: T. Fisher

Unwin, 1905,) p. 19. 'They [the Maasai] know that if they attempted to rise against the whites, all their old foes would be only too delighted to, in their turn, play the part of friendlies and, while helping the Government, get a little of their own back again.'

36 Leys to Harvey, No. 52, 29 Nov. 1914, Harvey Letters.

37 M. P. K. Sorrenson, *Origins of European Settlement in Kenya* (Nairobi: OUP, 1968), pp. 207, 126–7; Girouard to Harcourt, Tel. 217, 7 Oct. 1911, in response to CO to Girouard, 5 Oct. 1911, CO 533/91. For the ongoing row, see Bowring for Gov., No. 57, 29 Jan. 1913, CO 533/116; Bowker to Stordy, 23 Oct. 1912, in No. 57 and sequentially.

38 Smith, *Lily*, Ch. 31.

39 Affidavit of Stephano Ol-le Nongop, 25 June 1912, Enc. 1 in No. 75, Belfield to Harcourt, CO 879/112.

40 Leys to Harvey, Nos 26 (17 July 1911), 27 (20 July), 28 (31 July); Leys to Acting Gov. Bowring via the Principal Medical Officer, Enc. 1 in No. 51, 22 July 1912, and others in this register, CO 879/112.

41 In Enc. 3, No. 51, ibid.

42 Leys to Harvey, No. 27, ibid.

43 Bowring to Harcourt, 8 Aug. 1912, Conf. 80, Desp. 51, CO 879/112.

44 Morrison to Acting Chief Secretary, 4 Sept. 1912, (v) in Enc. 1 in No. 75, Belfield to Harcourt, CO 879/112; Enc. 12 in Conf. 136, 17 Dec. 1912, CO 533/109.

45 See references to Ole Sempele by R. Waller in T. Spear and I. N. Kimambo (eds), *East African Expressions of Christianity* (Oxford: James Currey), Ch. 5.

46 Morrison to CO, No. 60, 14 Sept. 1911, CO 879/112.

47 K. King, 'The Kenya Maasai and the protest phenomenon, 1900–1960', *JAH*, 12 (1971), pp. 120–31; Waller refers to Taki in Spear and Kimambo, *Expressions*.

48 Enc. 2 in No. 75, 13 Sept. 1912, CO 879/112; Enc. 3 (b) in No. 75, 5 Sept. 1912, CO 879/112; Enc. 4 in No. 75, 25 Sept. 1912, CO 879/112.

49 'A Govt. agent tried to terrify Legalishu once more, and refused to allow him to sell the cattle necessary to pay Mr M's fees. Mr M. applied to Nairobi and was refused', Leys to MacDonald, Harvey and Murray, No. 26, 17 July 1912. As for quarantine, 'The country into which Legalishu has been moved is in quarantine ... [it] was already gazetted infected with ECF ... He can't send a bullock out without breaking the law, so he cannot pay Mr Morrison's fees', Leys to Harvey, No. 29, 9 Sept. 1912.

50 Morrison to CO, 12 Aug. 1912, CO 879/112; 2 Oct. 1912, CO 869/112.

51 Hollis to Morrison, 7 Aug. 1912, (k) in Enc. 1, No. 75, CO 879/112; Acting Chief Secretary Monson to Home, 30 Sept. 1912, (aa) in Enc. 1, ibid.; Read to Anderson, 20 Jan. 1913, CO 533/109.

52 Leys to Harvey, No. 18, 22 Mar. 1911, Harvey Letters.

53 Leys to Harvey, No. 14, 24 Jan. 1911; Ross to mother, Nos 556, 565 (14 Aug. 1912, 29 Sept. 1911), McGregor Ross Papers; Conf. No. 80, Desp. 51, 8 Aug. 1912, CO 879/112; Leys to Harvey, Nos. 39 and 40, 30 Mar. and 10 April 1913, Harvey Letters.

54 Leys to Harvey, Nos 29 (9 Sept. 1912); 20 (12 May 1912); 23 (29 May 1912); 1 (17 Oct. 1910).

55 Leys to Harvey, No. 1, ibid.

56 Leys to Harvey, No. 1, ibid; Leys to Buxton, 10 June 1928, ASAPS Papers, box G143, RHO.
57 Interview with Veronica Bellers August 2002.
58 Leys to Harvey, Nos 21 and 22, 19 and 24 May 1912, Harvey Letters.
59 All letters cited are in Mss Harcourt Dep. 497 (Masai), Harcourt Papers, Bodleian Library.
60 G. H. Goldfinch Letters from Dec. 1923, box G137 in ASAPS Papers, RHO. Some of these are poorly catalogued. The Momonyot community, related to the Il-Laikipiak, managed to evade the 1911–13 move. For mention of this later move, and an analysis of the Leroghi land dispute which ended in Samburu victory over white settlers, see C. J. Duder and G. L. Simpson, 'Land and murder in colonial Kenya: The Leroghi land dispute and the Powys "murder" case', *Journal of Imperial and Commonwealth History*, 25, No. 3 (1997).
61 Statement of 22 Feb. 1923; undated 'Goldfinch's Notes on Native Administration'; undated paper 'Kenya as a White Colony'; undated paper, which appears to have been written in 1923 since it refers to 'the [Maasai] rebellion of last year', all in Goldfinch Letters, RHO; letter of 26 Aug. 1924 to Johnston, possibly from Buxton (no signature or name given), ASAPS Papers, box G143, RHO.
62 Talbot-Smith to PC Naivasha, 25 Sept. 1911, PC/RVP/6E/1/1, KNA.
63 Cashmore, 'Obedient', Vol. 1, pp. 129–30, he gives no source for the quote; Goldfinch to Buxton, 24 Nov. 1924, Goldfinch Letters, RHO.
64 K. Tidrick, *Empire and the English Character* (London: I. B. Tauris, 1990), pp. 177–9, 188.
65 Leys to Harvey, No. 15, 4 Mar. 1912, Harvey Letters.
66 'Senteu, Masai laibon', undated paper by Goldfinch, likely to be August 1925; also letters to Buxton, 29 Sept. and 6 Nov. 1925, Goldfinch Letters, RHO. Senteu's return from 'exile' is mentioned in the Narok Annual Report 1924–5. He died on 2 June 1933.
67 My thanks to Peter J. Ayre for finding additional information in antiquarian sources including Anthony Dyer, *Men for All Seasons: The Hunters and Pioneers* (Agoura, CA: Trophy Room Books, 1996). The others are not cited in the bibliography since I did not consult them directly.
68 Errol Trzebinski, *The Kenya Pioneers* (London: Heinemann, 1985), p. 139.
69 All the Gethin references are from 'An Old Settler Remembers Kenya', Richard Gethin Papers, RHO, pp. 25, 27; 23, 24, 30; and 7 of a separate, undated and untitled draft.

Part II The aftermath

Chapter 4 The court case

 1 Charles Eliot, *The East Africa Protectorate* (London: Edward Arnold, 1905), p. 197.
 2 'In the High Court of BEA at Mombasa, Civil Case No. 91 of 1912, *Ol le Njogo and Others* v. *The Attorney-General and Others*'. Plaint affirmed 25 February 1913 at Nairobi before E. R. Logan, Town Magistrate. The pleadings, plaint and related correspondence are in CO 533/116, printed as African No. 1001. Copies of the plaint are in the ASAPS Papers, G131, RHO. There is a 'Record

of the "Masai Case" ' in G. R. Sandford, *An Administrative and Political History of the Masai Reserve* (London: Waterlow & Sons, 1919), Appendix 3, pp. 186–222, which omits the May judgement.

3 Olonana's contested legacy is discussed in a short biography by Peter Ndege, *Olonana Ole Mbatian* (Nairobi: East African Educational Publishers, 2003) from p. 82.

4 These defendants were Ole Yele, Gilisho, Turere, Malit, Nakota, Batiet, Lingiri, Geeshen and Kotikall (several likely to be misspellings). That left Seggi, Ngaroya, Marmaroi, Saburi, Ayale, Ole Matipe, Ole Naigisa, Ole Tanyai and Masikonde opposing the action.

5 Sandford, *Administrative History*, pp. 190–1.

6 Plaint of Ol le Njogo (*sic*), *idem.*

7 Harvey raised this point in the Commons on 19 March 1913, *Hansard*, 5th series, Vol. 50, col. 1021.

8 His mother was Sinore, one of Ole Nchoko's five wives. He named the others as Nemardadi, Nemolel, Nampaiyio and Napelesh.

9 'Judgement of the High Court in the Case brought by the Masai Tribe against the Attorney-General of the East Africa Protectorate and Others; dated 26th May, 1913', Cd. 6939 (London: HMSO, 1913).

10 Ibid., p. 5.

11 Morrison to Harvey, No. 44, 18 June 1913, Harvey Letters.

12 Leys to Harvey, No. 45, 25 June 1913, Harvey Letters.

13 The *Nation*, BEA, 21 June 1913, pp. 447–8. Copy in the Harvey Letters. The original 'Naboth's Vineyard' was published in the *Nation* on 8 July 1911.

14 'Correspondence Relating to the Masai', Cd. 5584 (London: HMSO, June 1911).

15 Leys to Harvey, No. 45, 25 June 1913, Harvey Letters.

16 British East Africa – Its Legal Status', *Glasgow Herald*, 26 July 1913, no byline, copy in the Harvey Letters; Leys to Harvey, No. 47, 30 July 1913.

17 Note dated 22 July 1913, on desp. of 27 June, CO 533/119, ff499–500.

18 Minute by C. Tennyson on 'Masai claim to Laikipia: Pleadings in the High Court of BEA', 23 Jan. 1913, CO 533/129.

19 Frederick Jackson, *Early Days in East Africa* (London: Edward Arnold, 1930), pp. 330–1.

20 Minute by Harcourt on 'Masai claim to Laikipia', CO 533/129. On 28 January 1913 he wrote: 'I suppose if the case went against us in EAP we could take it to the Privy Council. Ought not the chiefs to be informed that they will have to share in the costs by sale of their cattle?'

21 Sandford, *Administrative History*, Appendix 3. Also *East African Law Reports* (EALR) 5, 1913, pp. 70–114. Only longer quotes from this judgement are page referenced here.

22 EALR 5, pp. 80–1.

23 Ibid., p. 97.

24 Ibid., pp. 110–11.

25 Y. P. Ghai and J. P. W. B. McAuslan, *Public Law and Political Change in Kenya* (Oxford: OUP, 1970), pp. 24–5.

26 EALR 5, J. Bonham-Carter, p. 99; J. King Farlow, p. 109.

27 Morrison to Leys, No. 50, 1 June 1914, Harvey Letters.

28 Morrison was owed 26 400 rupees, sued in August 1917 and lost in the High Court in January 1918, Cashmore, 'Obedient', p. 384.

29 Goldfinch to Travers Buxton, 19 Aug. (?1925), Goldfinch file, ASAPS Papers, RHO.
30 Leys, *Kenya*, pp. 114–15.
31 *KLC Evidence and Memoranda*, Vol. 3 (London: HMSO, 1934), p. 3267; original source R. L. Buell, *The Native Problem in Africa*, Vol. 1 (New York: Macmillan, 1928), p. 314.
32 The so-called Barth Judgement, EALR 102, 1921.
33 *KLC Evidence and Memoranda*, p. 3268.
34 M. F. Lindley, *The Acquisition and Government of Backward Territory in International Law* (London: Longmans, Green, 1926), pp. 321, 323, cited *KLC Evidence*, p. 3378.
35 *KLC Evidence*, p. 3379, citing F. D. Lugard, *The Dual Mandate in British Tropical Africa* (Edinburgh and London: William Blackwood & Sons, 1922), p. 288.
36 Claire Palley, 'The evolution of the powers of the Crown in protectorates', in Palley, *The Constitutional History and Law of Southern Rhodesia, 1888–1965: With Special Reference to Imperial Control* (Oxford: Clarendon Press, 1966).
37 Ghai and McAuslan, *Public Law*, pp. 22, 24.
38 David V. Williams, 'Unique Relationship Between Crown and Tangata Whenua?', in I. H. Hawharu (ed.), *Waitangi-Maori and Pakeha Perspectives of the Treaty of Waitangi* (Auckland: OUP, 1989), pp. 68–70.
39 Sandford, *Administrative History*, p. 36.

Chapter 5   The ecological impacts

1 The word means both human malaria and ECF in cows, see Glossary. My informants say it comes from the word *ntikan*, meaning swelling of the mandibular (lymphoid) tissue behind the ear, which is one of the first signs of ECF. It is a fatal disease of cattle caused by the protozoan parasite *Theileria parva*, carried largely by the brown ear tick *Rhipicephalus appendiculatus*. Even today, ECF is ranked as Africa's 'main livestock disease problem', killing 1.1m cattle each year. S. K. Mbogo, 'Trypanosomiasis and East Coast Fever', *Review of Kenyan Agricultural Research*, 40 (Bangor and Nairobi: University of Wales and KARI, 1998), pp. 17, 20.
2 Isaac Sindiga, 'Land and population problems in Kajiado and Narok, Kenya', *African Studies Review*, 27, No. 1 (March 1984), 27.
3 R. L. Tignor, *The Colonial Transformation of Kenya: The Kamba, Kikuyu and Maasai from 1900 to 1939* (Princeton and Guildford: Princeton University Press, 1976), p. 38.
4 Marcel Rutten, *Selling Wealth to Buy Poverty* (Saabrücken and Fort Lauderdale: Verlag Breitenbach, 1992), p. 8; D. Western and D. L. Manzolillo Nightingale, 'Environmental change and the vulnerability of pastoralists to drought: A case study of the Maasai in Amboseli, Kenya', in *Africa Environment Outlook Case Studies: Human Vulnerability to Environmental Change* (Nairobi: UNEP, 2004), pp. 31–50. For other examples, David J. Campbell, 'Response to drought among farmers and herders in southern Kajiado district, Kenya', *Human Ecology*, 12, No. 1 (1984), 35–64; M. Thompson and K. Homewood, 'Entrepreneurs, elites, and exclusion in Maasailand: Trends in wildlife conservation and pastoralist development', *Human Ecology*, 30, No. 1 (March 2002), 107–38.
5 M. Merker, *Die Mäsai* (Berlin: D. Reimer, 1904, 1910). In the Frieda Schütze translation, pp. 473–6.

6   R. D. Waller, ' "Clean" and "Dirty": Cattle disease and control policy in colo-
    nial Kenya, 1900–40', *JAH*, 45 (2004), 46. I wrote the chapter of my thesis
    upon which this chapter is based at least three years before this article was
    published, hence my ideas are not derived from it. He makes similar points
    about, for example, constructions of disease acting 'as a lens' through which
    to view social dynamics.

7   Ibid., p. 49; R. Waller and K. Homewood, 'Elders and experts: Contesting
    veterinary knowledge in a pastoral community', in A. Cunningham and
    B. Andrew (eds), *Western Medicine as Contested Knowledge* (Manchester:
    Manchester University Press, 1997). Informant Dickson Kaelo gave a vivid
    illustration of this. Agreeing with Waller and Homewood's claim that the
    'Maasai regarded disease as a natural, inevitable but potentially stable part of
    the environment' (p. 77), he described a saying Maasai use when branding
    cattle. The oldest man will urge a cow 'to be in the herd that collapses and
    grows – *tijinga naamuta eitu'*. In other words loss and gain are inseparable,
    and part of the natural cycle of life.

8   They were using it to prevent BPP by the second half of the nineteenth
    century, telling Merker that the method had been invented by the prophet
    Mbatiany. They also knew what caused malaria, and vaccinated against
    smallpox, Merker, *Die Mäsai*, pp. 224–6, 237–8.

9   These suggestions were made by Dickson Kaelo from interviews made on my
    behalf.

10 T. H. R. Cashmore, 'Your Obedient Servants' (see Bibliography), p. 343. He
    gives no source.

11 Dept. of Agriculture A/R 1911–12, p. 19. This recognised 'a degree of
    immunity existed in the cattle of certain districts', p. 18.

12 R. A. I. Norval, B. D. Perry and A. S. Young, *The Epidemiology of Theileriosis in
    Africa* (London: Academic Press, 1992), p. 51.

13 'Immunity in Native Cattle', Veterinary Pathologist A/R 1909–10, pp. 20–1.
    The Dept. of Agriculture A/R 1911–12, in a section on ECF that refers back to
    1906, stated: 'Even at this time it was suspected that a degree of immunity
    existed in the cattle of certain districts', p. 18.

14 Interview on Solio Ranch, Laikipia, January 2000. Bristow was then farm
    manager.

15 Drought refuges in the reserve, and major stock movements to and from
    them, are shown on Map 1 in E. A. Lewis, *A Study of the Ticks in Kenya Colony*,
    Part 3, 'Investigations into the tick problem in the Masai reserve', Bulletin
    No. 7 of 1934 (Nairobi: Government Printer, 1934).

16 Sindiga, 'Land and population problems', p. 27.

17 Norman Leys, *Kenya* (London: The Hogarth Press, 1924), p. 104.

18 K. M. Homewood and W. A. Rodgers, *Maasailand Ecology: Pastoralist
    Development and Wildlife Conservation in Ngorongoro, Tanzania* (Cambridge:
    Cambridge University Press, 1991) citing Waller: 1979. See also R. Lamprey
    and R. Waller, 'The Loita-Mara region in historical times: Patterns of subsis-
    tence, settlement and ecological change', Ch. 3 in Peter Robertshaw (ed.),
    Early Pastoralists of South-Western Kenya (Nairobi: British Institute in Eastern
    Africa, 1990).

19 T. R. McClanahan and T. P. Young (eds), *East African Ecosystems and their
    Conservation* (Oxford: OUP, 1996), pp. 225, 229.

20 D. E. Hutchins, *Report on the Forests of British East Africa* (London: HMSO, 1909); *Forests and Timber Resources of British East Africa*, no author given (London: Waterlow & Sons, 1920); R. S. Troup, *Report on Forestry in Kenya Colony* (London: Waterlow & Sons, for Government of Kenya, 1922), p. 10.

21 *Catalogue of the Forests of Kenya* (Forest Department, Kenya, August 1964), xeroxed copy in the Plant Sciences Library, Oxford.

22 My calculation of how much of Chepalungu lay inside the reserve is made on the basis of the map in the frontispiece of G. R. Sandford, *An Administrative and Political History of the Masai Reserve* (London: Waterlow & Sons, 1919), and E. A. Lewis, 'Tsetse-flies in the Ol Orokuti area of the Masai Reserve, Kenya Colony', *Bulletin of Entomological Research*, 28, Part 1 (1937). Lewis, who gave the size of Chepalungu as about 164 sq. miles, stated 'the greater part is outside ... the Masai boundary', p. 395.

23 Joseph Thomson, *Through Masai Land* (London: Sampson Low, Marston, Searle & Rivington, 1885), p. 405.

24 This was estimated higher in the 1930s, at 40–50 inches p.a., and the Southern Rift Valley at 10–30, Thos. E. Edwardson, *Regional Report on Kenya Colony* (unpublished, Imperial Forest Institute, Oxford, April 1934).

25 Thomson, *Through Masai Land*, pp. 407, 422.

26 Harry Johnston, *The Uganda Protectorate* (London: Hutchinson & Co., 1902), p. 2.

27 Charles Eliot, *The East Africa Protectorate* (London: Edward Arnold, 1905), pp. 80–2, 170–1.

28 Agricultural Report No. 2: 'Report on the Rift Valley District, Kijabe to Nakuru and Njoro', Enc. in No. 327, 9 May 1904, Eliot to Lansdowne, FO 2/836.

29 C. W. Hobley, 'Proposals as to the Procedure to be Adopted in Compensating Masai for Grazing Rights', an enc. in Desp. 495, 22 July 1904, FO 2/838.

30 'Journey from Naivasha to Baringo and the Laikipia Highlands', 24 June 1904, Enc. 1 in Desp. 493, 22 July 1904, FO 2/838.

31 W. Plowright, 'Inter-relationships between virus infections of game and domestic animals', *East African Agricultural and Forestry Journal*, 33 (June 1968), 262.

32 P. Robertshaw with R. Lamprey, in Robertshaw, *Early Pastoralists*, p. 13, describing in particular the western areas of Loita-Mara.

33 Masai A/R 1914–15, no page number legible on microfilm.

34 Robertshaw and Lamprey, in Robertshaw, *Early Pastoralists*, p. 11; Lamprey and Waller, 'The Loita-Mara Region', pp. 23, 25, latter citing E. A. Lewis, 'Tsetse flies and development in Kenya Colony', *East African Agricultural Journal*, 7 (1941–42).

35 Apart from an outbreak of trypanosomiasis on one farm in Naivasha district, the Veterinary Division A/R 1917–18 said no cases had been diagnosed in the highlands, p. 174. Government entomologists reported *G. pallidipes* at nearly 4000 feet – the highest point at which it had ever been found in BEA (Division of Entomology A/R year ending 31 Mar. 1918, p. 91). I am informed by Dr Glyn Davies, a retired VO who worked in Kenya, that Laikipia is beyond the range of *Glossina* transmission, but that the introduction of infected cattle and biting flies can set up local foci of transmission by mechanical means.

36 I shall not cover tsetse in any detail; it is another story. The early reports include R. B. Woosnam, 'Report on a search for *Glossina* on the Amala (Engabei) River, Southern Masai Reserve, East Africa Protectorate', *Bulletin of Entomological Research*, IV (1914), 272–8; and later E. A. Lewis, 'Tsetse flies in the Masai Reserve, Kenya Colony', *Bulletin of Entomological Research*, 25, Part 1 (1934), 439–55; Lewis, 'Tsetse flies and development in Kenya Colony', Part 1, *East African Agricultural Journal*, 7 (1941–42), 184–7. Also see the section on tsetse in Lamprey and Waller, 'The Loita-Mara Region', pp. 25–8; R. Waller, 'Tsetse fly in western Narok, Kenya', *JAH*, 31 (1990), 81–101, which briefly mentions the links between ticks and tsetse in the Lemek area, note p. 89.

37 Lewis, *A Study of the Ticks*, Part 3. The two previous parts of this study were: Part 1, 'A report on an investigation into the tick problem in the Rift Valley, Kenya Colony' (Bulletin No. 17 of 1931); Part 2, same title, ending in 'Rift Valley (contd.), the Uasin Gishu and Trans-Nzoia Districts, Kenya Colony' (Bulletin No. 6 of 1932).

38 Rumuruti Dist. A/R to March 1922.

39 *KLC Evidence*, Vol. 3 (a), p. 2306.

40 *Sidai*, pl. *sidain*, *sidan*, good, fine, nice, beautiful, handsome; *a-adua*, to be bitter, from *ol-odua* meaning gall-bladder, bitterness and rinderpest, Frans Mol, *Maasai Language and Culture Dictionary* (Limuru, Kenya: Kolbe Press, 1996), pp. 298, 370. Dickson Kaelo said his informants used *kemelok* interchangeably with *sidai*, particularly to refer to pastures; from the verb *a-melok*, 'to be sweet', Mol, ibid., p. 252.

41 *Hansard*, 5th series, Vol. 40, 1912 (June 24 to July 12), p. 475.

42 T. J. Anderson, Report of the Dept. of Agriculture, 1917–18, 89–92, cited in Lewis, 'Tsetse-flies', pp. 441–2. Lewis gives a useful précis of the official reports on the western extension of the reserve between 1908 and 1914, published and unpublished, only some of which I have covered here. Of the ten reports and other communications, seven mentioned the presence of tsetse, while two others reported other kinds of biting fly annoying to cattle, pp. 440–2. During military operations in World War I tsetse flies were collected and fly-belts located in the reserve, but many of these records were missing from departmental files, p. 442.

43 Vacated infected areas cleanse themselves of ticks after 15 months on average, Theiler and Stockman cited in Paul F. Cranefield, *Science and Empire: ECF in Rhodesia and the Transvaal* (Cambridge: Cambridge University Press, 1991), p. 186. Ticks are capable of transmitting infection for up to 15 months, D. J. Pratt and M. D. Gwynne (eds), *Rangeland Management and Ecology in East Africa* (London: Hodder & Stoughton, 1977), p. 184. The infection could have died out if no cattle had used this trail since the connecting road was closed in 1908, but herders, squatters and stock thieves may well have 'trespassed' there. The eastern move route was described as 'from east side of Olbolossat via Gilgil and Lake Naivasha' in the A/R of the Chief Stock Inspector 1912–13, p. 42. The only map I have been able to find of the proposed connecting road between the two reserves is one by Ainsworth in 1904, which shows it passing south-east of Lake Ol Bolossat, west of Kinangop, and straight down the eastern side of the Rift to Kedong. 'Rough sketch map prepared by Mr Ainsworth', in Enc. 3, Desp. 495, 22 July 1904, FO 2/838.

44 There are four stages in the life cycle of *R. appendiculatus*: egg, larva, nymph and adult. It is the nymph that attaches itself to an animal and transmits the disease. The adult fertilised female lays eggs within about six days of dropping to the ground; the larvae emerge 32 days later; moulting of larva and nymph takes 42 days in all. Lewis calculated that ticks spent at least 101 days on the ground, *A Study of the Ticks*, Part 2, p. 27.

45 These parasites have recently been shown to be genetically similar and are now classified as *T. parva* (cattle derived) and *T. parva* (buffalo derived), depending on the originating host, Mbogo, 'Trypanosomiasis and East Coast Fever', p. 17.

46 See study of BCT in Maasai cattle in the Loita Hills in N. Giles, F. G. Davies, W. P. H. Duffus and R. Heinonen, 'Bovine cerebral theileriosis', *The Veterinary Record*, 102, Issue 14 (April 8, 1978), 313. Also *ILRAD Reports*, 4, No. 2 (Nairobi: International Laboratory for Research on Animal Diseases, April 1986), on the breakdown of cattle immunity due to the wide antigenic diversity displayed by *T. p. lawrencei* parasites carried by buffalo, p. 3.

47 From the Dept. of Agriculture A/R 1911–12.

48 Report by Robert Stordy on stock diseases, Desp. 641, 29 Nov. 1905, CO 533/5. Altitude is obviously interconnected with other factors including soils, rainfall and humidity, which impact upon the life cycle of the tick.

49 See Dept. of Agriculture A/R 1912–13. Cattle were 'probably immune' in native reserves where ECF was endemic, p. 20. It considered the advisability of trying to produce 'a condition of endemicity' throughout the infected and partly infected areas of BEA similar to that in the reserves, p. 22. White farmers were encouraged to adopt the African practice of early exposure of calves, in order to build up an immune herd.

50 '*Engamuni*' was ascribed by the Maasai to 'cattle rubbing against trees on which rhinoceri have älso scratched themselves', Dept. of Agriculture A/R 1911–12. *M-benik* or *m'benek* is now known as ephemeral fever.

51 Laikipia District A/R year ending 31 March 1911, no page no. legible.

52 Ibid.

53 Stordy, Veterinary A/R 1912–13.

54 Acting Chief VO's A/R 1919–20, in the A/R of the Dept. of Agriculture.

55 'Report of the Convention of Associations', 25–29 October 1926, Nairobi, p. 122, quoting the Chief VO's report 1925. Col. Paterson of the Gilgil Farmers' Association told the meeting that his district had never had ECF before 1926, pp. 126–7.

56 'It was transport work with undipped grade bullocks in infected areas that spread the disease more than anything else. Transport cattle spread the disease all along the road,' Major Pardoe of the Molo Settlers Assn. told the Convention of Associations, p. 162.

57 Veterinary A/R 1912–13, quoted in Lewis, *A Study of the Ticks*, Part 3, p. 9.

58 Report of W. Kennedy, 21 Apr. 1913, in the Veterinary Dept. A/R 1912–13, pp. 51–2. The sheep disease is likely to have been sheep pox, or Nairobi sheep disease (NSD) which is transmitted by *R. appendiculatus*.

59 Norval, Perry and Young, *Epidemiology of Theileriosis*, p. 56.

60 Veterinary Division A/R 1917–18, p. 148.

61 Sandford, *Administrative History*, pp. 64–5.

62 Information supplied by Dickson Kaelo, from interviews with Ndeyo Ole Yiaile and Konana Kereto, and by Dr Nathan Ole Lengisugi. In BEA dips were first built along main roads in 1912 and farmers were encouraged to build their own. By 1916 there were 80 dips in the whole country, both private and state-run (Norval, Perry and Young, p. 54) and by 1923 there were 42 government tanks and 250 private ones. Conversely, Paul Mosley claims the first dip tanks were built in 1925, *The Settler Economies: Studies in the Economic History of Kenya and Southern Rhodesia, 1900–1963* (Cambridge: Cambridge University Press, 1983) note p. 244.

63 Leys, *Kenya*, note p. 105; *A Last Chance in Kenya* (London: Hogarth Press, 1931), p. 95.

64 A/R of the Acting Chief VO 1920, *idem*, pp. 20–1.

65 For examples from the Masai A/Rs, in 1915 it was said no VO had even visited the reserve all year except to procure meat supplies. In 1918, 'no conspicuous activity has been displayed by the Veterinary Department'. In 1920, the VO was removed after 'a certain amount of spasmodic activity'. In 1921, there was no VO in the reserve for seven months. In 1928–9, no VO visited Narok District all year. By 1933, there was still no VO there. It was only in 1938 that the A/R announced 'a year of activity and progress'.

66 From a debate on 31 January 1921, *Kenya Legislative Council Proceedings 1911–1923* (Nairobi: Government Printer), p. 64.

67 A/R of the Acting Chief VO 1920, *idem*, p. 20.

68 Chief VO Major Brassey-Edwards's reply to F. Ryder, who asked what the official rinderpest policy was in reserves. Stockowners' Conference organised by the Department of Agriculture, Nairobi, 10–11 March 1936 (Nairobi: Government Printer, 1937).

69 Waller, ' "Clean" and "Dirty" ', describes how post-war policy changed, pp. 63–4 *et seq.*

70 J. B. Orr and J. L. Gilks, *Studies of Nutrition: The Physique and Health of Two African Tribes*, Medical Research Council Studies of Nutrition, Special Report Series No. 155 (London: HMSO, 1931). They admitted having no health statistics for people living in remote parts of the reserve, relied over-much on examining patients and outpatients at the one hospital in Narok, where 20 per cent of patients were not Maasai, and said Maasai women probably did not seek hospital treatment as much as men did.

71 S. L. and H. Hinde, *The Last of the Masai* (London: Heinemann, 1901), p. 112.

72 William H. McNeill, *Plagues and Peoples* (New York: Anchor Press/Doubleday, 1976; London: Penguin 1994), pp. 27, 54, 89, 71.

73 Leys, *Kenya*, p. 283.

74 Lewis, *A Study of the Ticks*, Part 3, pp. 48–9. Sheep do not get ECF, but NSD is transmitted by the same tick.

75 Ibid., p. 60.

76 Ibid., p. 28.

77 Merker, *Die Mäsai*, p. 168 of the Schütze translation.

78 Lewis, *A Study of the Ticks*, Part 3, p. 53.

79 Sources are respectively: 'Native Administration – The Native Affairs Department and the Chief Native Commissioner', undated paper likely to have been written *c.* 1923–24, p. 17; Goldfinch to Buxton, 14 Feb. 1925;

covering letter, Goldfinch to Buxton, 5 Mar. 1923, enclosing statements by
Legeshaur (*sic*) and one other, pp. 2–3; Goldfinch to Buxton, 30 May 1924,
all Goldfinch Letters, ASAPS Papers, RHO.

80 R. W. Hemsted, 'Report on the Trans-Mara Extension of the Masai Reserve',
3 Apr. 1913, EAP No. 323, in Belfield to Harcourt, 5 May 1913, pp. 65–6,
CO 533/118, copy seen in the Harcourt Papers, Bodleian Library, Oxford. He
went on: 'I question if the area is suitable for the Masai cattle, or at all events
those from Laikipia ... If the Laikipia people occupied this area, they would
lose a large number of stock owing to the climatic conditions being quite dissimilar to anything they have been used to, even if they were not decimated
by East Coast Fever', pp. 66–7. For the history of Maasai sections in Trans-
Mara, see R. Waller, 'Interaction and identity on the periphery: The Trans-
Mara Maasai', *The International Journal of African Historical Studies*, 17, No. 2
(1984), 243–84.

Part III   Interpretations

Chapter 6   Blood oaths, boundaries and brothers

1 Elspeth Huxley, *White Man's Country: Lord Delamere and the Making of Kenya*,
Vol. 2 (London: Macmillan, 1935), p. 45, quoting Delamere's report, presumably to government, of a meeting on 24 September 1918; no original source
given.

2 Gerald Hanley, *Warriors and Strangers* (London: Hamish Hamilton, 1971),
p. 302.

3 Most other informants said the Maasai prophet involved was Olonana, son
of Mbatiany.

4 Meaning unclear. One informant said it meant 'rich in cows'. Elspeth Huxley,
who spelled it Nyasore, said it meant 'the lean man', *Out in the Midday Sun*
(London: Pimlico, 2000), p. 106.

5 Written communication.

6 From *ol-kiyieu*, brisket, Frans Mol, *Maasai Language and Culture Dictionary*
(Limuru, Kenya: Kolbe Press, 1996), p. 208.

7 Fig tree (*Ficus natalensis*): *o-reteti*, pl. *il-retet*, one of four trees holy to the
Maasai, ibid., p. 344. The tree at the alleged oath site was identified as *Ficus
Hochsetteri*.

8 Until his death in 2002, he lived near Entapipi, Naivasha, on land given to him
by Colvile. Huxley's informant Paddy Grattan called him Colvile's 'headman
cum ADC', Huxley Papers, RHO. His surname is derived from *ol-musunkui*, a
loanword from the Swahili *mzungu*, broadly meaning white person or
European, also a restless person or wanderer. He said he was called Swahili
because people believed he was fathered by a 'Swahili' (coastal) farm worker.

9 Unofficial members were unelected, nominated by the Governor. Delamere
was initially one of two unofficial members representing the settlers; the
other was Arthur Baillie.

10 Laikipia A/R 1910–11, p. 12, DC/LKA/1/1, KNA. Also Enc. 3 in Conf. Desp. 14,
CO 533 / 116.

11 E. Cole, *Random Recollections of a Pioneer Kenya Settler* (Woodbridge: Baron
Publishing, 1975), p. 44.

12 Personal communication. She pointed out that she was only 18 when her adoptive father died, and therefore has 'minimal knowledge of his relationship with the Maasai'. He never mentioned an oath.

13 F. D. Lugard, *The Rise of Our East African Empire: Early Efforts in Nysaland and Uganda* (Edinburgh: William Blackwood & Sons, 1893). Huxley describes how Delamere read avidly about African travel in the six months he spent flat on his back after a hunting accident, prior to first visiting East Africa, *White Man's Country*, Vol. 1, p. 25.

14 Reproduced in M. Perham (ed.), *The Diaries of Lord Lugard*, Vol. 1 (London: Faber & Faber, 1959), p. 421.

15 Errol Trzebinski describes some of these in her portrait of Denys Finch Hatton, *Silence Will Speak* (London: Heinemann, 1997), Ch. 3.

16 Evidence before the Commission at Narok, 19 October 1932, *KLC Evidence and Memoranda*, Vol. 2, p. 1198.

17 A. C. Hollis, *The Masai: Their Language and Folklore* (Oxford: Clarendon Press, 1905), p. 322.

18 C. W. Hobley, *Eastern Uganda: An Ethnological Survey* (London: Anthropological Institute of Great Britain and Ireland, 1902), p. 42; H. H. Johnston, *The Uganda Protectorate* (London: Hutchinson, 1902), p. 884; Hollis, *The Masai*, p. 322. Charles Eliot also mentioned the peace-making ceremony at Sangaruna, saying this was made between the agricultural and pastoral sections of the Maasai, *The East Africa Protectorate* (London: Edward Arnold, 1905), p. 142.

19 H. R. McClure, 'District Records for the Guidance of the Officer Administrating the Masai Southern Reserve', p. 12. Undated copy in the library of the British Institute in Eastern Africa, Nairobi, which appears to have been written before February 1910.

20 J. R. L. Macdonald, *Soldiering and Surveying in British East Africa, 1891–1894* (London: Edward Arnold, 1897), p. 35.

21 Diaries of Francis Hall, Hall Papers, RHO, p. 25 of the 1894 diary.

22 Harry Tegnaeus, *Blood-Brothers: An Ethno-Sociological Study of the Institutions of Blood-Brotherhood with Special Reference to Africa* (London and Stockholm: Kegan Paul, Trench, Trubner, Ethnographical Museum, 1952), p. 70.

23 Ibid., pp. 61–2.

24 M. Merker, *Die Mäsai* (Berlin: D. Reimer, 1904; 1910), pp. 70–72. Taken from an English translation of the 1910 edition in the Spiritan Library, Arusha. Translator and date are not given. My thanks to Dorothy Hodgson for sharing excerpts. Mol spells the word *ol-mumai*, pl. *il-muma*, meaning oath, *Maasai Language*, p. 263.

25 A. H. Neumann, *Elephant-Hunting in East Equatorial Africa* (London: Rowland Ward, 1898; Bulawayo: Books of Zimbabwe, 1982), pp. 42–4, 128, 130 of 1982 edn.

26 W. S. and K. Routledge, *With a Prehistoric People: The Akikuyu of British East Africa* (London: Edward Arnold, 1910), pp. 176–7.

27 Tegnaeus, *Blood-Brothers*, p. 501.

28 Ibid., p. 52; L. R. Von Höhnel, *Discovery of Lakes Rudolf and Stefanie: A Narrative of Count Samuel Teleki's Exploring and Hunting Expedition in East Equatorial Africa in 1887 and 1888* (London: Longmans, Green, 1894), from p. 314.

29 L. S. B. Leakey, *The Southern Kikuyu before 1903*, Vol. 1 (London: Academic Press, 1977), p. 60.

30   L. S. B. Leakey, *The Southern Kikuyu before 1903*, Vol. 1, pp. 59–60, 90–105, 491–6.

31   Diary entry for 13 Feb. 1894, Hall Papers, RHO, pp. 42–3.

32   R. D. Waller, 'The Lords of East Africa: The Maasai in the Mid-Nineteenth Century (*c.* 1840–1855)', Ph.D. thesis (Cambridge: 1978), p. 124. His sources are oral texts gathered by J. Gallagher.

33   Luise White, 'Blood brotherhood revisited: Kinship, relationship, and the body in East and Central Africa', *Africa*, 64, No. 3 (1994), 369.

34   E. Hertslet, *The Map of Africa by Treaty*, Vol. 1 (London: Frank Cass, 1967), lists 84 treaties made between 1887 and 1891 by agents of what became the IBEAC, pp. 374–8.

35   See Sir John Milner Gray, 'Early treaties in Uganda, 1888–91', *Uganda Journal*, Vol. 12 (March 1948), 29, 31; H. M. Stanley, *In Darkest Africa* (London: Sampson Low, Marston, Searle & Rivington, 1890), where references to blood-brotherhood include Vol. 1, pp. 358–61; Vol. 2, pp. 348–50. Expedition surgeon T. H. Parke wrote his own account of these ceremonies in *My Personal Experiences in Equatorial Africa: As Medical Officer of the Emin Pasha Relief Expedition* (London: Sampson, Low, Marston, 1891).

36   A. T. Matson, *Nandi Resistance to British Rule, 1890–1906* (Nairobi: East African Publishing House, 1972), p. 169.

37   Ibid., pp. 62, 66. In *Early Days in East Africa* (London: Edward Arnold), Jackson described making blood-brotherhood with chief Kaniri and two of his councillors near Athi Plains in 1889 (p. 170 of the 1930 edn); with unnamed Lumbwa (Kipsigis) in 1889 (p. 211); and with Kimangichi, chief of a community of former cave-dwellers on Mount Elgon, 1890 (p. 248).

38   C. C. Von der Decken and Otto Kersten, *Baron Carl Claus Von der Decken's Reisen in Ost-Afrika in den Jahren 1839 bis 1865* (Leipzig and Heidelberg: C. F. Winter, 1869–79), p. 309.

39   T. Ternan, *Some Experiences of an Old Bromsgrovian: Soldiering in Afghanistan, Egypt and Uganda* (Birmingham: Cornish Brothers, 1930), pp. 199–200; Matson, *Nandi Resistance*, pp. 89–90.

40   This last ceremony was held on 19 September 1891, Perham, *Lugard Diaries*, Vol. 2, p. 297. Stanley's claims to have made treaties with six chiefs were contested by Perham, *Lugard: The Years of Adventure, 1858–1898* (London: Collins, 1956), p. 259; and Gray, 'Early treaties', from p. 32. Gray said the dates and descriptions did not add up, and several 'treaties' were more likely to have consisted of blood-brotherhood alone.

41   The references are respectively, re-Ntale: Lugard, *The Rise*, Vol. 2, p. 160, and Perham, *Lugard*, Vol. 1, p. 262; re-Wakoli: Lugard, *The Rise*, Vol. 1, p. 369, and Perham, *Lugard*, Vol. 1, p. 416; re-Mbekirwa: Lugard, *The Rise*, Vol. 1, p. 370; and re-Kamba: ibid., p. 319.

42   Ibid., pp. 329–30.

43   Ibid., p. 330.

44   Captain F. D. Lugard, 'Treaty-making in Africa', *The Geographical Journal*, Vol. 1, No. 1 (January 1893), 53.

45   Leakey, *Southern Kikuyu*, Vol. 1, pp. 491–6. No dates.

46   'Masai and Kikuyu Swear the Ancient Peace Oath', *The East African Annual 1948–49* (Nairobi: *The East African Standard*, 1949), pp. 48–9.

47   White, 'Blood brotherhood revisited', p. 368.

48   Paul Spencer, *The Maasai of Matapato: A Study of Rituals of Rebellion* (Manchester: Manchester University Press, 1988), p. 264. See pp. 252–69 for

a full discussion of the significance of meat eating and particular cuts of meat. Cuts are ritually paired, and there is a deliberate emphasis on the opposition of human pairs in ceremonial – celebrants are linked to ritual partners, who are, for example, in turn 'opposed' to a pair of patrons who bless them. 'By pairing [people] with paired cuts of meat they remain opposed and yet are uniquely united', p. 261. The pairing of Delamere and Colvile with Ole Gilisho and Olonana, which is central to most stories about the blood-brotherhood, makes sense in this context.

Chapter 7   Highland games: settlers and their farm workers

1 See Elspeth Huxley, *White Man's Country: Lord Delamere and the Making of Kenya*, Vol. 1 (London: Macmillan, 1935), pp. 53–5. Her description of Laikipia, as shown through Delamere's eyes, was mouth-watering.

2 While leading the army's ninth division in March 1900, Colvile failed to relieve a column that had been ambushed by General De Wet. Lord Roberts, commander in chief of British field forces, accused him of 'reprehensible lack of vigour'. Two months later, he got into worse trouble after ignoring calls to help a Col. Spragge, whose men were surrounded and later decimated. Colvile was recalled to Britain, and pensioned off in January 1901. *Dictionary of National Biography*, Vol. 1 (Oxford: Oxford University Press), pp. 393–5. According to this, he died while riding a motorbike, but Huxley said it was a bicycle, *Out in the Midday Sun* (London: Pimlico, 2000), p.105.

3 Richard Gethin Papers, Mss. Afr. s. 1277 (1), RHO, f39.

4 Josslyn Hay, Earl of Erroll, was murdered near Nairobi in January 1941. The killer was never found. See James Fox, *White Mischief* (London: Penguin, 1984); and Errol Trzebinski, *The Life and Death of Lord Erroll* (London: Fourth Estate, 2000).

5 He was 55, Diana 25 years younger. Huxley describes the surprising courtship and marriage in *Midday Sun*, pp. 108–9. NB: Each generation of Delamere male heirs is called either Tom or Hugh; hence the current Baron is Hugh, his only son and heir Tom.

6 Trzebinski, *The Life and Death*, pp. 69, 112.

7 Huxley, *Midday Sun*, p. 105.

8 Oral testimony of Bristow, former manager of Colvile's Entapipi farm, and a former employee of the second Lord Delamere.

9 Huxley, *Midday Sun*, p. 107. The current Lord Delamere told me it took the family 20 years to get the farms back.

10 Kathryn Tidrick, 'The Masai and their Masters', Chapter 5 of Tidrick, *Empire and the English Character* (London: I. B. Tauris, 1990); 'The Masai and their Masters', *African Studies Review*, 23 (1980), 15–31.

11 Huxley, *White Man's Country*, Vol. 1, p. 39.

12 Personal communication.

13 Huxley, *White Man's Country*, Vol. 1, pp. 151–2.

14 Evidence of Lord Delamere to the Native Labour Commission 1912–13 (NLC), *Evidence and Report* (Nairobi: Government Printer, 1914), pp. 108–9.

15 G. R. Sandford, *An Administrative and Political History of the Masai Reserve* (London: Waterlow & Sons, 1919), p. 129; Narok District A/R 1925, KNA, p.19.

16 Interviewed at Soysambu, 2001.

17 Huxley, *White Man's Country*, Vol. 1, p. 138.

18   E. Cole, *Random Recollections of a Pioneer Kenya Settler* (Woodbridge: Baron Publishing, 1975), p. 46. The sheep were badly hit by heartwater in particular in 1919.

19   NLC, *Evidence and Report*: Hill, p. 112; Keeling, p. 112; Chaplin, p. 121; Delamere, from p. 108. Leys's written evidence on 26 December 1912 is pp. 270–4. McGregor Ross gave oral evidence, pp. 42–7.

20   The Narok District A/Rs, for example, do not give exact numbers of Maasai working outside the reserve. References are vague – 'a fair number of Laioni [boys] have gone out of the Reserve to work as herds' (1914–15); similar was said of warriors and boys in 1915–16; an estimated 400 Maasai working on stock farms, with a few employed as police trackers (1925); the two-year drought drove an 'increasing number' of young men out to find work, largely at Elmenteita and Laikipia because they refused to go anywhere else (1928).

21   Bruce Berman, *Control & Crisis in Colonial Kenya* (Oxford: James Currey, 1990), p. 62. Also J. Lonsdale in Berman and Lonsdale, *Unhappy Valley, Conflict in Kenya and Africa, Book 2: Violence and Ethnicity* (Oxford: James Currey, 1992); Tabitha Kanogo, *Squatters and the Roots of Mau Mau* (Oxford: James Currey, 1987).

22   Berman, *Control*, pp. 158, 219.

23   A. Clayton and D. C. Savage, *Government and Labour in Kenya, 1895–1963* (London: Frank Cass, 1974), p. 150. This percentage was much lower than that of other ethnic groups, e.g. 72 per cent of Lumbwa males were working, and 72.8 per cent of Kikuyu from Kiambu.

24   D. M. Anderson, 'Policing, prosecution and the law in colonial Kenya, c 1905–39', in D. M. Anderson and D. Killingray (eds), *Policing the Empire: Government, Authority and Control, 1830–1940* (Manchester: Manchester University Press, 1991).

25   Tabitha Kanogo, *Squatters and the Roots of Mau Mau* (Oxford: James Currey, 1987), p.13.

26   C. S. Nicholls, *Elspeth Huxley: A Biography* (London: HarperCollins, 2002).

27   'In mid-1931, a million and a half acres, 20 per cent of alienable Highland land [*sic*], had never been occupied by its owners [and] almost three million ... lay undeveloped by whites on existing occupied holdings', M. G. Redley, 'Politics of a Predicament: The White Community in Kenya, 1918–32', Ph.D. thesis (Cambridge: 1976), pp. 17–18. Cited in Berman, *Control*, p. 188n.

28   Ibid., p. 65.

29   Leys's written evidence to the NLC, *Evidence and Report*, p. 273.

30   Norman Leys, *Kenya* (London: The Hogarth Press, 1924), pp. 160–2.

31   Berman, *Control*, p. 129.

32   D. M. Anderson, 'Master and servant in colonial Kenya, 1895–1939', *JAH*, 41 (2000) describes the range of legislation affecting African workers in this period.

33   John Lonsdale, 'The moral economy of Mau Mau', in Berman and Lonsdale, *Unhappy Valley, Book 2*, pp. 383–4.

34   I am indebted to Desmond Bristow for permission to read and quote these. The diaries for 1927–29 and 1932 appear to have been written by a Captain B. W. D. Cochrane.

35   Leys, *Kenya*, p. 206. Wage rates were higher on the coast, and in towns.

36   NLC, *Evidence and Report*, p. 121.
37   Streptothricosis is a bacterial disease of cattle that causes dermatitis. Narok District A/R 1928, DC/NRK.1/1/2, KNA, p. 25.
38   A. Clayton and D. C. Savage, *Government and Labour in Kenya, 1895–1963* (London: Frank Cass, 1974), p. 151.
39   Berman, *Control*, pp. 57–8.
40   Leys, *Kenya*, p. 105.
41   'Probably [Colvile] would never have settled in East Africa had he not blown off several toes while shooting rabbits, which disqualified him from taking up the commission in his father's regiment that he had gained on leaving Sandhurst', Huxley, *Midday Sun*, p. 105.
42   Gethin Papers, RHO.
43   Delamere worked for Colvile for six months on graduating from Cambridge with an agricultural degree. His stepmother Diana had been married to Colvile for 12 years before their amicable split and her marriage to Tom Cholmondeley; Colvile was practically family. Incidentally, she is buried between the two on a small hill in the middle of the Entapipi plain, on what was Colvile's bull paddock. Nearby lies the grave of her and Colvile's only natural child, who died 10 days after birth. Bristow buried both Diana and Tom, and says the wall around the gravesite was Diana's idea: 'Colvile wanted cattle to walk freely over his grave.'
44   She did not use this information in *Midday Sun*. Huxley Papers, RHO.
45   See Tzebinski, *Kenya Pioneers*, Ch. 12. Settlers were taken on both as regular soldiers and intelligence scouts.
46   The Ross references are pp. 44–5 (ff98–9) of 1277 (1) and in Part IV of the Gethin Papers, RHO. These papers are disorganised and not properly catalogued.
47   Letter dated 22 January 1996 from Enaiborr-ajijik Group Ranch to then President Moi, re-claims to Land Parcel nos. 1382/1, 1382/2, 1378, 410/1, 410/2 and 6253 (Naivasha). Moi never replied; last I heard the claim was ongoing.
48   Huxley, *White Man's Country*, Vol. 2, p. 7. Trzebinski describes Delamere's role in *Kenya Pioneers*, from p. 180.
49   Sandford, *Administrative History*, from p. 126. The main assistance given was not in the form of conscripts, but in stock to feed the troops. Total stock provided by March 1917 from the Southern Reserve was around 300,000 sheep and 30,000 bullocks, of which about half the sheep and nearly all the bullocks were forcibly taken.
50   See *KLC Evidence and Memoranda* (London: HMSO, 1934), pp. 1199–200 for 'Dorobo' statements. Other reports say most stayed on in Laikipia and other northern areas, and were only removed later; e.g. the 'Makalia Wanderobo' living in the Nakuru Lake Forest Reserve were to have moved to Mau-Narok in October 1937, but all took jobs with Captain H. M. Harries and G. Lindstrom and moved on to their farms in order to avoid the forced move. Hislop to Tisdall, 25 Jan. 1938, Handing Over Report Nakuru-Ravine-Naivasha Districts, DC/LKI/1/1, KNA.
51   M. Merker, *Die Mäsai* (Berlin: D. Reimer, 1904, 1910), p. 70 of Spiritan translation.
52   R. D. Waller, 'The Lords of East Africa: The Maasai in the Mid-Nineteenth Century (*c.* 1840–1885)', Ph.D. thesis (Cambridge: 1978), pp. 291–2, 297;

Paul Spencer, *Nomads in Alliance: Symbiosis and Growth among the Rendille and Samburu of Kenya* (London: OUP, 1973), Appendix: The Dorobo and Elmolo of Northern Kenya, pp. 199–219. For Lee Cronk's published work, see Bibliography.

53 Laikipia A/R 1926, DC/LKI/1/1, KNA. Also see *KLC Evidence* and PC/RCP/6a/1/1/2 and 3, KNA.

54 'The Wanderobo or Hunting Tribe of Kenya', undated paper by Goldfinch in ASAPS Papers, RHO. By property, he means cows given in return for bringing in old ivory.

55 Huxley, *Midday Sun*, p. 109 of Pimlico edn (2000).

56 He means hounds. Colvile kept a pack of mixed breeds for hunting jackal, lion and leopard, Anne Carnelley to Huxley, Huxley Papers, RHO; Gethin Papers, 1227 (1), f40, RHO.

57 Goldfinch to Buxton, 1 Apr. 1924, Goldfinch Letters in the ASAPS Papers, RHO.

Conclusion

1 From an unpublished draft 'Concept Paper for the Facilitation of Activities towards Institution of the Maasai Case', by SIMOO (Simba Maasai Outreach Organisation), Kenya, 2003.

2 Part of my Conclusion has been published as an article, and will not be duplicated here. See 'Malice in Maasailand: The historical roots of current political struggles', *African Affairs*, 104/415 (April 2005), 207–24.

3 Remarks made by Ole Gilisho in a meeting at Nakuru, 5 Dec. 1912, Conf. 136, Belfield to Harcourt, 17 Dec. 1912, CO 533/109.

4 From a debate on the colonies, 20 July 1911, *Hansard*, Vol. 28, col. 1350.

5 Diana S. Wylie, 'Critics of Colonial Policy in Kenya with Special Reference to Norman Leys and W. McGregor Ross', M.Litt. thesis (Edinburgh: 1974), p. 65.

6 Memorandum on Masai Treaties of 1904 and 1911, 23 Mar. 1962, CO 822/2000.

7 See my dissertation, Ch. 4. This issue has been cut for reasons of length.

8 See my dissertation, Ch. 9. This has been dropped; my plan is to turn it into a journal article.

9 I am thinking in particular of the Hodgson-edited collection *Rethinking Pastoralism in Africa: Gender, Culture and the Myth of the Patriarchal Pastoralist* (Oxford: James Currey, 2000).

10 John W. Cell also noted this 'fundamental ambivalence', *By Kenya Possessed: The Correspondence of Norman Leys and J. H. Oldham, 1918–1926* (Chicago: University of Chicago Press, 1976), pp. 10–11.

11 Norman Leys, *Kenya* (London: The Hogarth Press, 1924), p. 392.

12 Cell, *By Kenya Possessed*, p. 8.

13 Hughes, *African Affairs*.

14 James Belich, *Making Peoples: A History of the New Zealanders from Polynesian Settlement to the End of the Nineteenth Century* (Auckland and London: Allen Lane, 1996), p. 195.

15 Vincent O'Malley, 'Treaty-making in colonial New Zealand', *New Zealand Journal of History*, Vol. 33, 2 (1999), 139.

16 Personal communication. Also quoted in my article, 'Fight for the forbidden land', *Sunday Times Magazine* (London: 4 January 2004).
17 Draft Declaration on the Rights of Indigenous Peoples, Part VI, Operative Paragraph 25, viewable at www.cwis.org/fwdp/International/draft9329.txt
18 ILO Convention 169 (1989), Convention Concerning Indigenous and Tribal Peoples in Independent Countries, viewable at www.cwis.org/fwdp/International/ilo_169.txt
19 *UK Foreign and Commonwealth Office Human Rights Annual Report 2004* (London: 2004), p. 212.
20 Chris Mullin MP, Parliamentary Under Secretary of State, Foreign and Commonwealth Office, to Lord Avebury, 23 Mar. 2005.

Bibliography

This does not include titles of works not cited in the main text, newspaper sources, or items not directly consulted. See chapter notes for other references.

Manuscript and archival sources

In Britain

The Bodleian and its dependent libraries, including Rhodes House (RHO) and Radcliffe Science (RSL), Oxford

Anti-Slavery and Aborigines' Protection Society Papers, including the G. H. Goldfinch Letters, RHO.
The Anti-Slavery Reporter and Aborigines' Friend, Vols 1–4 (1909–1914), RHO.
Autobiography of Alfred Claud Hollis, unpublished, RHO.
Blue Books for EAP, RHO (Nairobi: Government Printer).
Clarence Buxton Papers, RHO.
Catalogue of the Forests of Kenya, no author given (Forest Department, August 1964), Plant Sciences Library, Oxford.
Robert Chamberlain Papers, RHO.
Report of the Convention of Associations, 25–29 October 1926 (Nairobi), RHO.
Delamere, Lord, 'Letters from Kenya', Mss. Afr. s. 1424(1) RHO.
East African Annual 1948–49 (Nairobi: East African Standard, 1949), RHO.
Richard Gethin Papers, RHO.
Francis Hall Papers, RHO.
Lord Claud Hamilton Papers, RHO.
Lewis Harcourt Papers, Bodleian.
Elspeth Huxley Papers, RHO.
Johnston, H. H., 'Report on the Nyasa-Tanganyika Expedition 1889–90', in the British South Africa Co. Papers, RHO.
Journal of the East Africa and Uganda Natural History Society, Vols 1–6 (1910–1917) and July 1919 (No. 14, no vol. given), RSL.
Kenya Legislative Council Proceedings, 1911–23 (Nairobi: Government Printer), RHO.
Gilbert Murray Papers, Bodleian.
McGregor Ross Papers, RHO.
Native Labour Commission 1912–13, Evidence and Report (Nairobi: Government Printer, 1914), RHO.
Parliamentary Debates (*Hansard*), 5th Series, Vols 26–8, 30, 34, 37, 40–1, 47, 50, 53, 54, 56, Bodleian.
Proceedings of the Royal Geographical Society, Vols 5–8, for reports of nineteenth-century exploration by Farler, Fischer, Last, Thomson and Von der Decken, Bodleian.
Reports of Stockowners' Conferences in 1927, 1930, 1936 (Nairobi: Government Printer), RHO.

The Veterinarian, Vols LXXIV–V, Nos. 882–3 (June 1901, Jan. 1902), RSL.
The Veterinary Record, Vols 22–3 (1910–11), RSL.

Public Record Office, London

Foreign and Colonial Office archives for BEA, largely series CO 533, 628, 879, FO 2 and 107. Some items listed under the title 'Printed primary sources' were also seen here.

In Kenya

British Institute in Eastern Africa Library, Nairobi

McClure, H. R., 'District Records for the Guidance of the Officer Administrating the Masai Southern Reserve', undated copy.

ILRI Library, Nairobi

ILRAD Reports, various issues (1986–92) of quarterly newsletter of the International Laboratory for Research on Animal Diseases, now the International Livestock Research Institute (ILRI).

Libraries at Kabete Veterinary Laboratories and Muguga (formerly the East African Veterinary Research Organisation), near Nairobi

East African Agricultural Journal, Vols 9 and 16 (October 1943, October 1950).
Veterinary Dept. A/Rs from 1908 (and as a division within the Dept. of Agriculture); A/Rs of the Veterinary Pathologist, Chief Stock Inspector, Quarantine Officer. Some also viewed at RHO and the KNA.

Kenya National Archives, Nairobi

(Much of this is also available on microfilm at RHO.)
Dept. of Agriculture A/Rs, BEA, from 1911.
Game Dept. A/Rs 1925–6.
Handing Over Report, Nakuru-Ravine-Naivasha Districts, 1938, DC/LKI/1/1.
Laikipia District Survey of Events 1906–11; Laikipia Quarterly and Annual Reports 1909–27.
Land files including LND 30/3/4/5 and LND 30/3/4/5/19 (re-Magadi Soda Co.).
Machakos A/Rs, correspondence and Political Records, various from 1899.
Masai A/Rs 1914–39; occasional later reports to 1947.
Masai Move 1911: Reports from officers in charge (PC/RVP/6E/1/1); related correspondence in PC/RVP/6E/1/3 and 1/5.
Murumbi Papers, MAC/KEN/100/4–6.
Nandi District Quarterly Reports 1909.
Narok District A/Rs 1914–40.
Southern Masai Reserve District Records 1908–11.
Ukamba Province Files 1906, 1912–15.

Miscellaneous

Bellers, Veronica, 'What Mr Sanders Really Did', unpublished draft mss.
The Harvey Letters (largely the letters of Norman Leys to Edmund Harvey, 1910–14), private collection on loan to this author. To be deposited in a public archive.

Diaries of Lariak Farm Manager, 1927–29, 1932. Held by Desmond Bristow.

Mol, Frans, unpublished notes on the life of Parsaloi Ole Gilisho.

'Notes on Andrew "Trader" Dick', supplied to author by great-nephew J. A. Dawson.

Smith, Michael L., *The Lily and the Rose*, unpublished biography of Percy Girouard.

Written communications with Veronica Bellers, Dick Cashmore, Arthur Cole, Deborah Colvile, Lee Cronk, Glyn Davies, Dorothy Hodgson, Frans Mol, Walter Plowright.

Printed primary sources

'Correspondence Relating to the Masai', Cd. 5584 (London: HMSO, June 1911).

'Further Correspondence Relating to the Masai', 17 July 1911–20 July 1914, African No. 1001 (London: HMSO, 1915).

'Correspondence Relating to the Resignation of Sir Charles Eliot, and to the Concession to the East Africa Syndicate', Cd. 2099, Africa No. 8 (London: HMSO, July 1904).

Eliot, Charles, 'Report on the East Africa Protectorate', Cd. 769 (London: HMSO, 1901).

'Judgement of the High Court in the Case brought by the Masai Tribe against the Attorney-General of the East Africa Protectorate and Others; dated 26th May, 1913', Cd. 6939 (London: HMSO, 1913).

Kenya Land Commission Evidence and Memoranda, 3 vols (London: HMSO, 1934).

'Masai Enquiry Committee Report' (Nairobi: East African Standard, February 1926).

Native Labour Commission 1912–13, Evidence and Report (Nairobi: Government Printer, 1914).

'*Ol le Njogo and Others* v. *The Attorney General and Others*', Court of Appeal for Eastern Africa, 7 Dec. 1913, *East African Law Reports*, 5, 70–114.

Printed secondary works

1   Books

Anderson, D. M., 'Policing, prosecution and the law in colonial Kenya, c. 1905–39', in Anderson and Killingray, D. (eds), *Policing the Empire: Government, Authority and Control, 1830–1940* (Manchester: Manchester University Press, 1991).

—— and Grove, R. (eds), *Conservation in Africa: People, Policies and Practice* (Cambridge: Cambridge University Press, 1987).

—— and Johnson, D. H. (eds), *Revealing Prophets* (Oxford: James Currey, 1995).

Belich, James, *Making Peoples* (Auckland and London: Allen Lane, 1996).

Berman, Bruce, *Control & Crisis in Colonial Kenya: The Dialectic of Domination* (Oxford: James Currey, 1990).

Berman, B. and Lonsdale, J., *Unhappy Valley: Conflict in Kenya & Africa, Books 1 and 2* (Oxford: James Currey, 1992).

Brockington, D., *Fortress Conservation: The Preservation of the Mkomazi Game Reserve, Tanzania* (Oxford: James Currey, 2002).

Buell, R. L., *The Native Problem in Africa*, Vol. 1 (New York: Macmillan, 1928).

Cell, John W. (ed.), *By Kenya Possessed: The Correspondence of Norman Leys and J. H. Oldham, 1918–1926* (Chicago: University of Chicago Press, 1976).

Chevenix-Trench, C., *Men Who Ruled Kenya: The Kenya Administration, 1892–1963* (London: Radcliffe Press, 1993).

Churchill, Winston, *My African Journey* (London: Hodder & Stoughton, 1908).

Clayton, A., and Savage, D. C., *Government and Labour in Kenya, 1895–1963* (London: Frank Cass, 1974).

Cole, E., *Random Recollections of a Pioneer Kenya Settler* (Woodbridge: Baron Publishing, 1975).

Collett, D., 'Pastoralists and wildlife: Image and reality in Kenya Maasailand', in Anderson and Grove, *Conservation in Africa* (1987).

Cranworth, Lord, *A Colony in the Making* (London: Macmillan, 1912).

Cronk, Lee, *From Mukogodo to Maasai: Ethnicity and Cultural Change in Kenya* (Boulder, CO: Westview Press, 2004).

Decken, C. C. Von der, and Kersten, Otto, *Baron Carl Claus Von der Decken's Reisen in Ost-Afrika in den Jahren 1839 bis 1865* (Leipzig and Heidelberg: C. F. Winter, 1869–79).

Dening, Greg, *Performances* (Chicago: University of Chicago Press, 1996).

Eliot, Charles, *The East Africa Protectorate* (London: Edward Arnold, 1905).

Galaty, J., 'Maasai pastoral ideology and change', in Salzman, P. C. (ed.), *Contemporary Nomadic and Pastoral Peoples: Africa and Latin America* (Williamsburg, VA: Department of Anthropology, College of William and Mary, 1982).

——, 'Pastoral orbits and deadly jousts: Factors in the Maasai expansion', in Galaty and Bonte, P. (eds), *Herders, Warriors and Traders* (Boulder, CO: Westview Press, 1991).

Ghai, Y. P. and McAuslan, J. P. W. B., *Public Law and Political Change in Kenya* (Nairobi: OUP, 1970).

Goldsmith, F. H., *John Ainsworth: Pioneer Kenya Administrator, 1864–1946* (London: Macmillan, 1955).

Hanley, Gerald, *Warriors and Strangers* (London: Hamish Hamilton, 1971).

Hill, M. F., *Permanent Way: The Story of the Kenya and Uganda Railway* (Nairobi: East African Railways and Harbours, 1950).

——, *Magadi: The Story of the Magadi Soda Company* (Birmingham: The Kynoch Press for the Magadi Soda Co.), 1964.

Hinde, H. and Hinde, S. L., *The Last of the Masai* (London: Heinemann, 1901).

Hindlip, Lord Charles Allsop, *British East Africa: Past, Present and Future* (London: T. Fisher Unwin, 1905).

Hobley, C. W., *Kenya: From Chartered Company to Crown Colony* (London: H. R. and G. Witherby, 1929; 2nd edn Frank Cass, 1970).

Hochschild, Adam, *King Leopold's Ghost* (London: Macmillan, 1999).

Hodgson, Dorothy L. (ed.,) *Rethinking Pastoralism in Africa: Gender, Culture and the Myth of the Patriarchal Pastoralist* (Oxford: James Currey, 2000).

——, 'Images and interventions: The problems of pastoralist development', in Anderson and Broch-Due (eds), *The Poor Are Not Us* (Oxford: James Currey, 1999).

Höhnel, L. R. Von, *Discovery of Lakes Rudolf and Stefanie* (London: Longmans, Green, 1894).

Hollis, A. C., *The Masai: Their Language and Folklore* (Oxford: Clarendon Press, 1905).
——, *The Nandi: Their Language and Folklore* (Oxford: James Currey, 1969 edn).
Homewood, K. M., and Rodgers, W. A., *Maasailand Ecology: Pastoralist Development and Wildlife Conservation in Ngorongoro, Tanzania* (Cambridge: Cambridge University Press, 1991).
Hughes, L., ' "Beautiful beasts" and brave warriors: The longevity of a Maasai stereotype', in G. de Vos, L. Romanucci-Ross and T. Tsuda (eds), *Ethnic Identity: Problems and Prospects for the Twenty-first Century* (Walnut Creek, CA: AltaMira Press, 2006).
Hutchins, D. E., *Report on the Forests of British East Africa* (London: HMSO, 1909).
Huxley, Elspeth, *White Man's Country: Lord Delamere and the Making of Kenya*, 2 vols (London: Macmillan, 1935).
——, *Out in the Midday Sun* (London: Chatto & Windus, 1985; Pimlico, 2000).
—— and Perham, M., *Race and Politics in Kenya: A Correspondence between Elspeth Huxley and Margery Perham* (London: Faber & Faber, 1944).
Jackson, Frederick, *Early Days in East Africa* (London: Edward Arnold, 1930).
Jacobs, A. H., 'A chronology of the pastoral Masai', in Ogot, B. A. (ed.), *Hadith 1* (Nairobi: East African Publishing House, 1968).
——, 'Maasai pastoralism in historical perspective', in Monod, T. (ed.), *Pastoralism in Tropical Africa* (Oxford and London: OUP, 1975).
Johnson, D. H., and Anderson, D. M. (eds), *The Ecology of Survival* (Boulder, CO: Westview Press, 1988).
Johnston, H. H., *The Kilima-Njaro Expedition* (London: Kegan Paul, Trench, 1886).
——, *Britain Across the Seas* (London: National Society's Depository, 1910).
——, *The Uganda Protectorate* (London: Hutchinson, 1902).
Kanogo, Tabitha, *Squatters and the Roots of Mau Mau* (Oxford: James Currey, 1987).
King, K. and Salim, A. (eds), *Kenya Historical Biographies* (Nairobi: East African Publishing House, 1971).
Kipury, Naomi, *Oral Literature of the Maasai* (Nairobi: Heinemann, 1983).
Krapf, J. L., *Travels, Researches and Missionary Labours during Eighteen Years Residence in Eastern Africa* (London: Trübner, 1860).
Lamprey, R. and Waller, R., 'The Loita-Mara region in historical times: Patterns of subsistence, settlement and ecological change', in Robertshaw, *Early Pastoralists* (1990).
Leakey, L. S. B., *The Southern Kikuyu before 1903* (London: Academic Press, 1977).
Leys, Norman, *The Colour Bar in East Africa* (London: The Hogarth Press, 1941).
——, *A Last Chance in Kenya* (London: The Hogarth Press, 1931).
——, *Kenya* (London: The Hogarth Press, 1924).
Lugard, F. D., *The Rise of our East African Empire* (Edinburgh: William Blackwood & Sons, 1893).
Macdonald, J. R. L., *Soldiering and Surveying in British East Africa, 1891–1894* (London: Edward Arnold, 1897; Folkstone: Dawsons of Pall Mall, 1973).
Matson, A. T., *Nandi Resistance to British Rule, 1890–1906* (Nairobi: East African Publishing House, 1972).
Maxon, R. M., *Struggle for Kenya: The Loss and Reassertion of Imperial Initiative, 1912–1923* (London: Associated University Presses, 1993).
McClure, H. R., *Land-Travel and Seafaring* (London: Hutchinson, 1925).
McGregor Ross, W., *Kenya from Within* (London: George Allen & Unwin, 1927).

McNeill, W. H., *Plagues and Peoples* (New York: Anchor Press/Doubleday, 1976; London: Penguin, 1994).

Meinertzhagen, Richard, *Kenya Diary, 1902–1906* (Edinburgh: Oliver & Boyd, 1957).

Merker, M., *Die Mäsai* (Berlin: Dietrich Reimer, 1904; 1910).

Miller, Charles, *The Lunatic Express* (New York: Macdonald, 1971).

Mol, Frans, *Maasai Language and Culture Dictionary* (Limuru, Kenya: Kolbe Press, 1996).

Mungeam, G. H., *British Rule in Kenya, 1885–1912* (Oxford: Clarendon Press, 1966).

——, (ed.), *Select Historical Documents, 1884–1923* (Nairobi: East African Publishing House, 1978).

Muriuki, G., *A History of the Kikuyu, 1500–1900* (Nairobi: OUP, 1974).

Ndege, Peter, *Olonana Ole Mbatian* (Nairobi: East African Educational Publishers, 2003).

Neumann, A., *Elephant-Hunting in East Equatorial Africa* (London: Rowland Ward, 1898; Bulawayo: Books of Zimbabwe, 1982).

Nevinson, H. W., *A Modern Slavery* (London and New York: Harper, 1906).

Nicholls, C. S., *Elspeth Huxley: A Biography* (London: HarperCollins, 2002).

North, S. J., *Europeans in British Administered East Africa, 1889–1903* (Wantage: published by the author, 1995).

Norval, R. A. I., Perry, B. D. and Young, A. S., *The Epidemiology of Theileriosis in Africa* (London: Academic Press, 1992).

Oliver, Roland, *Sir Harry Johnston and the Scramble for Africa* (London: Chatto & Windus, 1957).

Orr, J.B. and Gilks, J.L., *Studies of Nutrition: The Physique and Health of two African tribes*, Medical Research Council Studies of Nutrition, Special Report Series No. 155 (London: HMSO, 1931).

Palley, Claire, *The Constitutional History and Law of Southern Rhodesia 1888–1965* (Oxford: Clarendon Press, 1966).

Perham, M., *East African Journey: Kenya and Tanganyika 1929–1930* (London: Faber & Faber, 1976).

—— (ed.), *The Diaries of Lord Lugard* (London: Faber & Faber, 1959).

——, *Lugard: The Years of Adventure 1858–1898* (London: Collins, 1956).

Portal, Sir G., *The British Mission to Uganda in 1893* (London: Edward Arnold, 1894).

Rigby, P., *Persistent Pastoralists: Nomadic Societies in Transition* (London: Zed Books, 1985).

Robertshaw, Peter, *Early Pastoralists of South-Western Kenya* (Nairobi: British Institute in Eastern Africa, 1990).

Routledge, W. S. and K., *With a Prehistoric People: The Akikuyu of British East Africa* (London: Edward Arnold, 1910).

Rutten, Marcel, *Selling Wealth to Buy Poverty: The Process of the Individualization of Landownership Among the Maasai Pastoralists of Kajiado District, Kenya, 1890–1990* (Saabrücken and Fort Lauderdale, FL: Verlag Breitenbach, 1992).

Saitoti, Tepilet Ole, *Maasai* (London: Elm Tree Books/Hamish Hamilton, 1980).

Sandford, G. R., *An Administrative and Political History of the Masai Reserve* (London: Waterlow & Sons, 1919).

Scott, James, *Domination and the Arts of Resistance: Hidden Transcripts* (New Haven, CT: Yale University Press, 1990).

Sorrenson, M. P. K., *Origins of European Settlement in Kenya* (Nairobi: OUP, 1968).

Spear, T. and Kimambo, I. N. (eds), *East African Expressions of Christianity* (Oxford: James Currey, 1999).

Spear, T. and Waller, R. (eds), *Being Maasai: Ethnicity and Identity in East Africa* (London: John Currey, 1993).

Spencer, Ian R. G., 'Pastoralism and colonial policy in Kenya, 1895–1929', in Rotberg, R. (ed.), *Imperialism, Colonialism and Hunger in East and Central Africa* (Lexington, MA: D. C. Heath, 1983).

Spencer, Paul, *The Maasai of Matapato: A Study of Rituals of Rebellion* (Manchester: Manchester University Press, 1988).

——, *Nomads in Alliance: Symbiosis and Growth among the Rendille and Samburu of Kenya* (London: OUP, 1973).

Stanley, H. M., *In Darkest Africa* (London: Sampson Low, Marston, Searle & Rivington, 1890).

Tegnaeus, H., *Blood-Brothers: An Ethno-Sociological Study of the Institutions of Blood-Brotherhood with Special Reference to Africa* (London and Stockholm: Kegan Paul, Trench, Trubner, Ethnographical Museum, 1952).

Ternan, T., *Some Experiences of an Old Bromsgrovian* (Birmingham: Cornish Brothers, 1930).

Thomson, Joseph, *Through Masai Land* (London: Sampson Low, Marston, Searle & Rivington, 1885).

Tidrick, K., *Empire and the English Character* (London: I. B. Tauris, 1990).

Tignor, R. L., *The Colonial Transformation of Kenya: The Kamba, Kikuyu and Maasai from 1900 to 1939* (Princeton, NJ, and Guildford: Princeton University Press, 1976).

Tonkin, E., *Narrating our Pasts: The Social Construction of Oral History* (Cambridge: Cambridge University Press, 1992).

——, 'History and the myth of realism', in Samuel, R., and Thompson, P. (eds) *The Myths We Live By* (London: Routledge, 1990).

Troup, R. S., *Report on Forestry in Kenya Colony* (London: Waterlow & Sons, 1922).

Trzebinski, Errol, *The Life and Death of Lord Erroll* (London: Fourth Estate, 2000).

——, *The Kenya Pioneers* (London: Heinemann, 1985).

——, *Silence Will Speak* (London: Heinemann, 1977).

UK Foreign and Commonwealth Office Human Rights Annual Report 2004 (London: 2004).

Vansina, Jan, *Oral Tradition as History* (Oxford: James Currey, 1997).

Waller, R., 'They do the dictating and we must submit: The Africa Inland Mission in Maasailand', in Spear and Kimambo, *East African Expressions of Christianity* (1999).

——, 'Kidongoi's kin: Prophecy & power in Maasailand', in Anderson and Johnson, *Revealing Prophets* (1995).

——, 'Emutai: Crisis and response in Maasailand, 1883–1902', in Johnson and Anderson, *The Ecology of Survival* (1988).

—— and Homewood, K., 'Elders and experts: Contesting veterinary knowledge in a pastoral community', in A. Cunningham and B. Andrews (eds), *Western Medicine as Contested Knowledge* (Manchester: Manchester University Press, 1997).

Western, D., and Nightingale, D. L. Manzolillo, 'Environmental change and the vulnerability of pastoralists to drought: A case study of the Maasai in Amboseli, Kenya', in *Africa Environment Outlook Case Studies: Human Vulnerability to Environment Change* (Nairobi: UNEP, 2004).

Williams, D. V., 'Unique Relationship Between Crown and Tangata Whenua?', in Hawharu, I. H. (ed.), *Waitangi-Maori and Pakeha Perspectives of the Treaty of Waitangi* (Auckland: OUP, 1989).

Wrigley, C. C., Ch. V in Harlow, V. and Chilver, E. M. (eds), *History of East Africa*, Vol. 2 (Oxford: Clarendon Press, 1965).

Zwanenberg, R. M. A. Van, with Anne King, *An Economic History of Kenya and Uganda, 1800–1970* (London and Basingstoke: Macmillan, 1975).

2   Articles and pamphlets

Anderson, D. M., 'Master and servant in colonial Kenya, 1895–1939', *JAH*, 41 (2000), 459–85.

——, 'Stock theft and moral economy in colonial Kenya', *Africa*, 56 (1986), 399–416.

Berntsen, John L., 'The enemy is us: Eponymy in the historiography of the Maasai', *History in Africa*, 7 (1980), 1–21.

——, 'Maasai age-sets and prophetic leadership, 1850–1912', *Africa*, 49, No. 2 (1979), 134–46.

Cronk, Lee, 'From true Dorobo to Mukogodo Maasai: Contested identity in Kenya', *Ethnology*, 41, No. 1 (2002), 27–49.

Fosbrooke, H. A., 'The life of Justin: An African autobiography translated and annotated by H. A. Fosbrooke', *Tanganyika Notes and Records*, Vols 41–2 (1955–56), 30–56, 19–29.

——, 'An administrative survey of the Masai social system', *Tanganyika Notes and Records*, No. 26 (December 1948), 1–50.

Gray, Sir J. M., 'Early treaties in Uganda, 1888–91', *Uganda Journal*, 12 (March 1948), 25–42.

Hodgson, Dorothy L., 'Pastoralism, patriarchy and history: Changing gender relations among Maasai in Tanganyika, 1890–1940', *JAH*, 40, No. 1 (1999), 41–65.

Hughes, Lotte, 'Malice in Maasailand: The historical roots of current political struggles', *African Affairs*, 104/415 (April 2005), 207–24.

Johnston, H. H., 'The East African problem', *The Nineteenth Century and After*, 380 (October 1908), 567–87.

'Kenya: White Man's Country? Report to the Fabian Colonial Bureau', pamphlet, no author given (London: Fabian Publications, 1944).

King, K., 'The Kenya Maasai and the protest phenomenon, 1900–1960', *JAH*, 12 (1971), 117–37.

Knowles, J. N., and Collett, D. P., 'Towards a historical understanding of development and conservation in Kenyan Maasailand', *Africa*, 59, No. 4 (1989), 433–59.

Lewis, E. A., 'Tsetse flies and development in Kenya colony', Part 1, *East African Agricultural Journal*, 7 (1941–42), 184–7.

——, 'Tsetse flies in the Ol Orokuti area of the Masai Reserve, Kenya colony', *Bulletin of Entomological Research*, 28, Part 1 (1937), 395–402.

——, 'A study of the ticks in Kenya colony', Parts 1–3, Bulletins No. 17 of 1931, No. 6 of 1932, No. 7 of 1934 (Nairobi: Government Printer).

——, 'Tsetse flies in the Masai Reserve, Kenya colony', *Bulletin of Entomological Research*, 25, Part 1 (1934), 439–55.

Lugard, F. D., 'Treaty-making in Africa', *The Geographical Journal*, 1, No. 1 (January 1893), 53–5.

Maxon, R. M., 'Judgement on a colonial governor: Sir Percy Girouard in Kenya', *Transafrican Journal of History*, 18 (1989), 90–100.

——, and Javersak, D., 'The Kedong Massacre and the Dick Affair: A problem in the early colonial historiography of East Africa', *History in Africa*, 8 (1981), 261–9.

Mungeam, G. H., 'Masai and Kikuyu responses to the establishment of British administration in the East Africa Protectorate', *JAH*, 11, No. 1 (1970), 127–43.

O'Malley, Vincent, 'Treaty-making in early colonial New Zealand', *New Zealand Journal of History*, 33, No. 2 (1999), 137–54.

Percival, A. B., 'Game and disease', *Journal of the East Africa and Uganda Natural History Society*, No. 13 (November 1918), 302–15.

Sindiga, Isaac, 'Land and population problems in Kajiado and Narok, Kenya', *African Studies Review*, 27, No. 1 (March 1984), 23–39.

Spencer, Paul, 'Opposing streams and the gerontocratic ladder: Two models of age organisation in East Africa', *Man*, 11 (1976), 153–73.

Tidrick, K., 'The Masai and their masters', *African Studies Review*, 23 (1980), 15–31.

Tignor, R. L., 'The Maasai warriors: Pattern maintenance and violence in colonial Kenya', *JAH*, 13 (1972), 271–90.

Waller, R., ' "Clean" and "dirty": Cattle disease and control policy in colonial Kenya, 1900–40', *JAH*, 45 (2004), 45–80.

——, 'Tsetse fly in western Narok, Kenya', *JAH*, 31 (1990), 81–101.

——, 'The Maasai and the British, 1895–1905: The origins of an alliance', *JAH*, 17, No. 4 (1976), 529–53.

——, 'Interaction and identity on the periphery: The Trans-Mara Maasai', *International Journal of African Historical Studies*, 17 (1984), 243–84.

White, Luise, 'Blood brotherhood revisited: Kinship, relationship, and the body in East and Central Africa', *Africa*, 64, No. 3 (1994), 359–72.

Woosnam, R. B., 'Report on a search for Glossina on the Amala (Engabei) River, Southern Maasai Reserve, E.A.P.', *Bulletin of Entomological Research*, 4 (1914), 271–8.

Wylie, Diana, 'Confrontation over Kenya: The colonial office and its critics, 1918–1940', *JAH*, 18, No. 3 (1977), 427–47.

——, 'A Debate on Empire', a review of J. W. Cell, *By Kenya Possessed*, *JAH*, 18, No. 4 (1977), 633–4.

——, 'Norman Leys and McGregor Ross: A case study in the conscience of African empire, 1900–39', *Journal of Imperial and Commonwealth History*, 5, No. 3 (May 1997), 294–309.

Other unpublished papers, dissertations and theses

Cashmore, T. H. R., 'Your Obedient Servants: Studies in District Administration in the East African Protectorate, 1895–1918' (1964), an earlier draft of what became a Ph.D. thesis (Cambridge: 1965). Available at RHO (Mss. Afr. s. 1034).

——, 'The Masai Succession Dispute', draft written *c.* 1969–70.

Edwardson, T. E., 'Regional report on Kenya colony', Imperial Forest Institute (Oxford: 1934).

Hughes, Lotte, 'Moving the Maasai: A Colonial Misadventure', D.Phil. thesis (Oxford: 2002).

Jacobs, Alan, 'The Traditional Political Organisation of the Pastoral Masai', D.Phil. thesis (Oxford: 1965).

——, 'The pastoral Masai of Kenya: A report of anthropological field research', paper for the Ministry of Overseas Development (London: 1963).

SIMOO (Simba Maasai Outreach Organisation), 'Concept Paper for the Facilitation of Activities toward Institution of the Maasai Case' (Kenya: 2002).

Waller, R. D., 'The Lords of East Africa: The Maasai in the Mid-Nineteenth Century (*c.* 1840–1885)', Ph.D. thesis (Cambridge: 1978).

Wylie, Diana S., 'Critics of Colonial Policy in Kenya, with Special Reference to Norman Leys and W. McGregor Ross', M.Litt. thesis (Edinburgh: 1974).

Index

Note: All informants, written sources and persons are not listed, only the most important or most substantively quoted. With one exception, chapter endnotes are not indexed.

230   *Index*